Gardens and Human Agency in the Anthropocene

This volume discusses gardens as designed landscapes of mediation between nature and culture, embodying different levels of human control over wilderness, defining specific rules for this confrontation and staging different forms of human dominance.

The contributing authors focus on ways of rethinking the garden and its role in contemporary society, using it as a crossover platform between nature, science and technology. Drawing upon their diverse fields of research, including History of Science and Technology, Environmental Studies, Gardens and Landscape Studies, Urban Studies, and Visual and Artistic Studies, the authors unveil various entanglements woven in the past between nature and culture, and probe the potential of alternative epistemologies to escape the predicament of fatalistic dystopias that often revolve around the Anthropocene debate.

This book will be of great interest to those studying environmental and landscape history, the history of science and technology, historical geography, and the environmental humanities.

Maria Paula Diogo is Full Professor of History of Technology and Coordinator of the Interuniversity Centre for the History of Science and Technology (CIUHCT), School of Sciences and Technology, NOVA University of Lisbon, Portugal.

Ana Simões is Full Professor of History of Science, Co-Coordinator of the Interuniversity Centre for the History of Science and Technology (CIUHCT), School of Sciences, University of Lisbon, Portugal, and President of the European Society for the History of Science.

Ana Duarte Rodrigues is Research Fellow of the Interuniversity Centre for the History of Science and Technology (CIUHCT), School of Sciences, University of Lisbon, Portugal.

Davide Scarso is a Post-Doc Researcher at the Interuniversity Centre for the History of Science and Technology (CIUHCT), School of Sciences and Technology, NOVA University of Lisbon, Portugal.

Routledge Environmental Humanities

Series editors: Scott Slovic (University of Idaho, USA), Joni Adamson (Arizona State University, USA) and Yuki Masami (Kanazawa University, Japan)

The *Routledge Environmental Humanities* series is an original and inspiring venture recognising that today's world agricultural and water crises, ocean pollution and resource depletion, global warming from greenhouse gases, urban sprawl, overpopulation, food insecurity and environmental justice are all *crises of culture*.

The reality of understanding and finding adaptive solutions to our present and future environmental challenges has shifted the epicentre of environmental studies away from an exclusively scientific and technological framework to one that depends on the human-focused disciplines and ideas of the humanities and allied social sciences.

We thus welcome book proposals from all humanities and social sciences disciplines for an inclusive and interdisciplinary series. We favour manuscripts aimed at an international readership and written in a lively and accessible style. The readership comprises scholars and students from the humanities and social sciences and thoughtful readers concerned about the human dimensions of environmental change.

Gardens and Human Agency in the Anthropocene

Edited by Maria Paula Diogo,
Ana Simões, Ana Duarte Rodrigues
and Davide Scarso

LONDON AND NEW YORK

First published 2019
by Routledge
2 Park Square, Milton Park, Abingdon, Oxon OX14 4RN

and by Routledge
605 Third Avenue, New York, NY 10017

First issued in paperback 2020

Routledge is an imprint of the Taylor & Francis Group, an informa business

British Library Cataloguing-in-Publication Data
A catalogue record for this book is available from the British Library

Library of Congress Cataloging-in-Publication Data
Names: Diogo, Maria Paula, editor.
Title: Gardens and human agency in the anthropocene / edited by Maria Paula Diogo, Ana Duarte Rodrigues, Ana Simäoes and Davide Scarso.
Description: Abingdon, Oxon ; New York, NY : Routledge, 2019. |
Series: Routledge environmental humanities | Includes bibliographical references and index.
Identifiers: LCCN 2018061191 (print) | LCCN 2019009656 (ebook) |
ISBN 9781351170246 (eBook) | ISBN 9780815346661 (hbk)
Subjects: LCSH: Gardens–Environmental aspects. | Gardens–History. |
Human ecology. | Landscape gardening–Environmental aspects.
Classification: LCC SB454.3.E53 (ebook) | LCC SB454.3.E53 G37 2019
(print) | DDC 635–dc23
LC record available at https://lccn.loc.gov/2018061191

ISBN: 978-0-367-72978-3 (pbk)
ISBN: 978-0-8153-4666-1 (hbk)

Typeset in Goudy
by Wearset Ltd, Boldon, Tyne and Wear

Contents

Illustrations

Figures

Table

Contributors

Vanessa Cirkel-Bartelt studied History, English and Medieval English Literature at Heinrich Heine University, Düsseldorf. She worked for the universities in Dortmund and Wuppertal, co-organizing interdisciplinary workshops and coordinating the project on the epistemology of the LHC. In 2011 she defended her PhD thesis about the history of early cosmic ray studies at the University of Wuppertal published in 2013. Since then she has worked on the history of energy; especially the public reception of scientific knowledge. Currently, she works on the British Atomic Gardening Movement and its implications for knowledge production in negotiation between laypeople and experts.

Maria Paula Diogo is Full Professor of History of Technology (FCT-NOVA) and Coordinator of the Interuniversity Centre for the History of Science and Technology (CIUHCT). She publishes, coordinates and participates in research projects, and organizes meetings on a regular basis both nationally and internationally. She currently leads the project "Anthropolands – Engineering the Anthropocene. The role of colonial Science, Technology and Medicine on changing of the African landscape". Recent publications include *Sciences in the Universities of Europe, Nineteenth and Twentieth Centuries. Academic Landscapes* (Springer, 2015), *Europeans Globalizing: Mapping, Exploiting, Exchanging* (Palgrave/Macmillan, 2016) and "STEP Forum", *Technology and Culture*, 57(4), 2016, 926–981.

Ana Duarte Rodrigues is an Assistant Professor of History of Science (FCUL) and researcher at the Interuniversity Centre for the History of Sciences and Technology. She is editor-in-chief of *Gardens & Landscapes* journal, published by Sciendo. She is the principal investigator of the research projects "Sustainable Beauty for Algarvean Gardens" (2015–2020) and "Horto Aquam Salutarem" (2018–2021), both funded by the Portuguese Foundation for Science and Technology. Her research is focused on gardens and landscapes through the perspective of the History of Science.

Aline de Figueirôa Silva is an architect and urbanist with a Master's degree in Urban Development from UFPE and a PhD in Architecture and Urbanism from USP. She is currently Assistant Professor at the School of Architecture of UFBA, Brazil, and conducts research on Salvador's historic gardens. At UFPE, she engaged in garden studies and the restoration of Burle Marx's gardens and was based at the Landscape Laboratory (2001–2010). She worked for the Brazilian National Institute of Artistic and Historic Heritage (2005–2010). In 2013, she was a Junior Fellow at Dumbarton Oaks Research Library and Collection/Harvard University. She is the author of *Jardins do Recife: uma história do paisagismo no Brasil, 1872–1937* (CEPE, 2010) and co-editor of *Jardins de Burle Marx no Nordeste do Brasil* (UFPE, 2013).

Johan Gärdebo is a PhD candidate in the History of Science, Technology and Environment at the KTH Royal Institute of Technology. Gärdebo's dissertation analyses how Swedish satellite remote sensing contributed to the making of the environment between 1969 and 2001. He holds two Bachelor's degrees in the Humanities and the Social Sciences from Uppsala University. He has published on the relationship between satellite remote sensing and orbital debris in *Anthropocene Review* and recently finished an oral history project on Swedish space activities in the late twentieth century. He has been a visiting graduate student at MIT (Massachusetts, USA) and NTNU (Norway).

inhabitants is an online channel for exploratory video and documentary reporting, launched in 2015 in New York by the visual artist **Mariana Silva** and the visual artist and writer **Pedro Neves Marques**. inhabitants produces and streams short-form videos intended for online distribution. It has collaborated with institutions such as Haus der Kulturen Der Welt and the Max Planck Institute for the History of Science (Berlin), the Berado Collection Museum (Lisbon), and is currently developing a video series for Contour8 (Mechelen, Belgium), as well as with artists such as Filipa César and Louis Henderson.

Luke Keogh is a historian and curator, and is currently Senior Curator at the National Wool Museum, Geelong, Australia. From 2012 to 2014 he worked at the Deutsches Museum, Munich, and was one of the curators on the exhibition *Welcome to the Anthropocene. The Earth in Our Hands* (2014–2016). His research and writing has received many awards, including the Redmond Barry Award from the State Library of Victoria and the Sargent Award from the Arnold Arboretum of Harvard University. His latest book is *The Wardian Case* (University of Chicago Press, 2019).

Eric LoPresti holds a BA in Cognitive Science (University of Rochester) and a MFA (Maryland Institute College of Art). His artwork examines the imposition of technology upon the environment and the aftermath of the Cold War. His dramatic landscapes juxtapose abstract elements with representations of the vast deserts of the American west, exploring relationships between science, identity, history and conflict. Recent solo exhibitions include *An Ocean of Light* at Burning in Water (New York City) and *Test Site* at the National Atomic Test Museum (Las Vegas). His work has received mentions in *New York Times* and *Nature*, and he has been interviewed on international video by Reuters and the *Washington Post*.

Ivo Louro holds a MSc degree in Environmental Engineering from the School of Sciences and Technology of the NOVA University of Lisbon. He has worked on Sustainable Public Procurement projects in the National Laboratory of Energy and Geology (LNEG). He has experience with sustainable public procurement, life cycle assessment and carbon footprints of products and services. He authored the thesis *Environmental Dimensions of Advertising. Case Study: LCA of an Advertising Campaign in the MUPI Format*, and co-authored the book *Sustainable Public Procurement: Training Guide for Public Authorities*. He is currently a Research Fellow at the Interuniversity Centre for the History of Science and Technology (CIUHCT, FCT-NOVA). He has become increasingly interested in the Anthropocene debate, namely in the questions of nature–culture dichotomy, unequal ecological exchange, planetary boundaries and safe operating space.

Nina Möllers holds a PhD in Modern History and is curator and manager of programmes and events for BIOTOPIA, a new type of museum for the life sciences currently being planned in Munich. As researcher and curator at Deutsches Museum, she headed the exhibition *Welcome to the Anthropocene. The Earth in Our Hands* (2014–2016). Her research interests are museum studies, environmental and gender history, and human–earth relations. Recent publications include *Objects in Motion. Globalizing Technology* (co-edited with Bryan Dewalt) (Smithsonian, 2016) and "Materializing the Medium. Staging the Age of Humans in the Exhibition Space", in *Zeitschrift für Medien- und Kulturforschung* (2018).

Ana Cristina Roque holds a PhD in the History of Discoveries and Expansion. She is Assistant Director of the CH-ULisboa (Center of History – University of Lisbon) and a member of the research group Building and Connecting Empires. She worked at the University Eduardo Mondlane, Mozambique (1983–1985), and at the Tropical Research Institute, Lisbon (1995–2015), where she integrated and coordinated different research projects concerning the CPLP countries and co-organized several international events. She works mainly on the History of Africa and the Indian Ocean, especially Mozambique (sixteenth to nineteenth centuries), favouring a multidisciplinary approach with particular emphasis on scientific expeditions, indigenous knowledge, health policies and environmental problems in a colonial context.

Davide Scarso is a post-doc researcher at the Interuniversity Centre for the History of Science and Technology (CIUHCT, FCT-NOVA), with a project on the philosophical, institutional and political challenges of the Anthropocene debate. Between 2015 and 2016 he lectured on Philosophy and Ethics at the Pontifícia Universidade Católica do Paraná (Curitiba, Brazil). From 2009 to 2014 he worked as a post-doctoral researcher at CIUHCT (project title: "The

Nature/Culture Polarity in Contemporary Anthropological Theories"). He obtained a PhD in the Philosophy of Science in 2008 from the Universidade de Lisboa (Portugal).

Astrid Schwarz is Full Professor of Technoscience Studies at Brandenburg University of Technology Cottbus-Senftenberg, Germany. She is interested in the identification and entwinement of theoretical and practical knowledge forms. Her philosophical field research covers scientific, social and artistic experimentation, with a focus on assemblages of technology and environment. Accordingly, she publishes on the making of technoscientific objects, on modes of imaging and agentic power. Recently, she carried out research on so-called gardening practices in the Anthropocene and the corresponding agent Homo hortensis. Her most recent books are *Experiments in Practice* (2014) and *Research Objects in their Technological Setting* (2017) co-authored with Bernadette Bensaude-Vincent, Sacha Loeve and Alfred Nordmann.

Ana Simões is Full Professor of History of Science (FCUL), Co-Coordinator of the Interuniversity Centre for the History of Science and Technology (CIUHCT), and President of the European Society for the History of Science. She has authored and edited more than 130 publications, participates in research projects and networks, and regularly organizes meetings, both nationally and internationally. She currently leads the project "Visions of Lisbon. Science, Technology and Medicine and the making of a technoscientific capital". Recent publications include *Neither Physics nor Chemistry. A History of Quantum Chemistry* (MIT Press, 2011, reprinted 2014) and "STEP Forum", *Technology and Culture*, 57(4), 2016, 926–981.

Ana Matilde Sousa holds a MSc degree in Fine Arts/Painting with the thesis *Bukimi-tan: On the Concept of Kawaii in Contemporary Japanese Art Production (An Itinerary)* from the School of Fine Arts of the University of Lisbon (FBAUL) and is currently pursuing a PhD in Painting at the same institution. She is affiliated with the Centre for Research and Studies in Fine Arts (CIEBA), and conducts research on contemporary Japanese art and popular culture. She has participated in group exhibitions since 2003 and is a founding member of the zine label Clube do Inferno, under which she has self-published her works and collaborated with projects both in Portugal and internationally since 2012. She has contributed regularly to the blog *L'Obéissance Est Morte* since 2013.

Helmuth Trischler is Head of Research at the Deutsches Museum, Munich, Professor of Modern History at Ludwig Maximilian University, Munich, and Director of the Rachel Carson Center for Environment and Society. His main research interests are innovation cultures in international comparison; science, technology and European integration; and environmental history. Jointly with Nina Möllers, he conceptualized the exhibition *Welcome to the Anthropocene. The Earth in Our Hands*, shown at the Deutsches Museum from 2014 to 2016. His recent books include *Building Europe on Expertise*.

Innovators, Organizers, Networkers (Palgrave/Macmillan, 2014), and *Cycling and Recycling. Histories of Sustainable Practices* (Berghahn, 2016).

Jaume Valentines-Álvarez holds a BE in Industrial Engineering, a BA in History and a PhD in the History of Science. His main current interests are the entanglement of nationalism, technocracy and crisis in Southern Europe, the politics of urban nuclear reactors during the Cold War, and the role of fun in anti-nuclear movements as an antidote to scientific ignorance. He is a research fellow at the CIUHCT (NOVA University of Lisbon), and has recently been a visiting scholar at the Max Planck Institute for the History of Science and the University of Geneva. He regularly participates with West Coast art collective in cross-discipline projects in Lisbon.

Nina Wormbs is Associate Professor at the Division of History of Science, Technology and Environment at KTH Royal Institute of Technology, Stockholm. She was its Chair From 2010 to 2016. She recently edited the volume *Competing Arctic Futures* (Palgrave, 2018). Her earlier work focused on media and technological change, especially broadcasting and the use of the spectrum, and her later work concerned Arctic assessments as well as remote sensing and the planetary. She is currently embarking on a project studying the legitimization of non-action in climate change.

Introduction

Nature and gardens in the history of science and technology and in garden and landscape studies

Maria Paula Diogo, Ana Simões, Ana Duarte Rodrigues and Davide Scarso

For centuries, science and technology have envisaged the relationship between humans and nature as unidirectional, nature being the passive protagonist serving as background for human agency. Doing justice to its dominant role, humankind has controlled nature through science, technology or other categories of knowledge, and as far as Western worldviews are concerned this asymmetric relationship has been at the core of the concept of progress and growth, which has framed Western societies.

When it comes to reflections about science and technology, history and philosophy of science and technology, among other disciplinary fields, have been questioning the absence of nature as a historical actor and narrative leader up until the twentieth century. Since the *École des Annales*, the 1960s debate on the culture–nature divide, and more recent scholarship on the entanglements between nature, technology and humans, have been under discussion (Bloch 1931, Febvre 1935, Braudel 1949, Kranzberg and Pursell 1967, Benjamin, 1986, Schatzki 2003, Williams 2010). They have been joined by the recent debates fostered by the concept of Anthropocene, which offer new avenues to explore nature's voice in the realm of history.

Starting from criticisms of the a-historicity often embedded in many debates on the Anthropocene, which use "we" and "us" as though human society is and has always been a homogeneous, flat and free-floating reality, they have called attention to alternative concepts to describe the "age of humankind", particularly by stressing the role played by capitalism, colonialism, imperialism and other forms of worldwide economy in the unbridled exploitation of natural resources. This perspective has been behind proposals to take a *long-durée* approach to deconstruct how a global technoscientific epistemology, based on the concepts of unlimited progress and growth, has converged to the *naturalization* of technology and the *commodification* of nature as structural building blocks of the hegemonic worldview which came to be associated with the age of the Anthropocene. Based on this hegemonic view, nature, as part of a larger domination-rooted type of society, became a human technoscientific construction (or destruction, for that matter), populated by mechanical bees, plastic trees that soak up carbon dioxide, artificial islands, climate geoengineering, cloned animals and cyborgs, to name a few. For some, this paradox is now perceived as the "new normality"; for others, starting with the environmental scientist and journalist Bill McKibben, the dramatic erosion of the divide between nature and (human technoscientific) culture entailed the end of nature as an autonomous category: "we have deprived nature of its independence, and this is fatal to its meaning. Nature's independence *is its meaning* – without it there is nothing but us" (McKibben 1989, 58).

Simultaneously, gardens and landscapes have become part of the concerns of historians of science and technology, most particularly since the so-called spatial turn (Livingstone 2003, 2010). By calling attention to the situatedness of technoscientific knowledge,[1] in the sense that "scientific knowledge is a geographical phenomenon. It is acquired in specific sites; it circulates from location to location; it transforms the world" (Livingstone 2010, 18), Livingstone has urged us to take seriously the role of "landscape agency", and proposes to assess "the role of landscape in the generation and circulation of scientific knowledge claims" in the loose sense of allowing "space for the thought that nature has some part to play in the theories that are constructed about it" (Livingstone 2010, 9). In this context, gardens and landscapes emerged as particularly rich topics for further analysis. Two examples of scholarship in the field are Matteo Valleriani and Jochen Büttner's project on the Garden of Pratolino,[2] and Antoine Picon's chapter on the park of Buttes-Chaumont (Picon 2010). The Garden of Pratolino is one of the several private gardens of Tuscany that contributed to the renaissance of ancient pneumatics, thus bringing together traditional theories and new practical knowledge; the Haussmanian park of Buttes-Chaumont, built during the French Second Empire, seems closely akin to wilderness, but it is, in fact, a heavy user of technology, not only above ground, but also below surface (sewers, shoring). The Buttes-Chaumont park, with all its technological devices, is in itself a "machine à produire de l'urbanité" (Picon 2010, 30) that was built to serve the specific ideological commitment of the Parisian bourgeoisie. In both cases, the scientific and technical apparatus behind the gardens, as well as the agency that underlies their construction, are brought to the forefront, unveiling strong ideological, political and social entanglements.

But gardens have traditionally been approached in the disciplinary context of gardens and landscape studies. By exploring their relationship with landscape architecture, the study of their cultural and artistic significance has brought together social and cultural history, geography, aesthetics, technology (including horticulture) and conservation. More recently, scholarship in this field has bridged traditional research on garden and landscape studies with topics of the history of science and technology, as exemplified by the recently published volume on *Gardens as Laboratories. The History of Botany through the History of Gardens* (Baldassarri and Matei 2017). There is, therefore, a vast literature on gardens and landscapes both from the perspective of visual and artistic studies and from the standpoint of the history of science and technology. There is also a growing literature on the Anthropocene and anthropogenic landscapes mostly from a scientific perspective or from eco-activism. We assess this trend in the following section.

Nature and gardens in the Anthropocene

While it could be said that announcements of the death of nature were probably "greatly exaggerated" it is not uncommon to see our present predicament described by sentences that include the term "doomed" (Hillman 2018, Malm 2018, Scranton 2018). Well before the word "Anthropocene" started its unstoppable march to popularity (according to Google trends, somewhere in the middle of 2011), in his celebrated *The End of Nature*, McKibben (1989) warned that human beings had become the "most powerful source for change on the planet" and explored the possible ethical implications of this awareness. He acknowledged, however, that declaring that nature as something separated and independent from human affairs has ended possessed a "clanging finality" that may give even more strength to the drive to control nature, through technics of genetic engineering or plans of planetary management. McKibben chose to advocate *restraint* as the real challenge, and argued that only by limiting "now, *today* [...] our numbers and our desires and our ambitions" may we have the chance to dismantle the greenhouse we have built "where once there bloomed a sweet and wild garden" (McKibben 1989, 78).

Two years later, and seemingly as a response to McKibben, Michael Pollan argued that the "all or nothing" logic of wilderness ethic – whose adherents "are apt to throw up their hands in despair and declare the 'end of nature'" (Pollan 1991/2003, 194) – would hardly be helpful in most contemporary environmental issues, as they are irremediably ambiguous. Most often than not, the choice is not whether to intervene or not, but to be able to distinguish different degrees, methods and objectives of intervention, just as gardeners do. A gardener, according to Pollan, is a down-to-earth practical individual who "doesn't waste too much time in metaphysics" (227) trying to establish the ideal approach to nature, and "tends not to be romantic" (192) about it. As his/her knowledge comes from direct experience with natural processes, s/he does not picture himself/herself as working from the outside and, therefore, s/he "is not likely to conclude from the fact that some intervention in nature is unavoidable, therefore 'anything goes'" (Pollan 1991/2003, 194). Consequently, an environmental ethics that has gardening as its central metaphor and model allows for an open and frank discussion regarding the aim and methods of environmental interventions. Despite an unavoidable anthropocentric character, gardens foster a "give-and-take" and adaptable approach that escapes both a mystifying cult of wilderness and a biotechnology-driven hubris:

> Gardens [...] teach the necessary if un-American lesson that nature and culture can be compromised, that there might be some middle ground between the lawn and the forest, between those who would complete the

conquest of the planet in the name of progress, and those who believe it is time we abdicate our rule and leave the earth in the care of its more inno-cent species.

(Pollan 1991/2003, 64)

At the time, in academic circuits, these issues were being discussed under the label of a "social construction of nature" (see Proctor (1998) and Hacking (1999) for early critical appraisals). At the end of the 1980s, with the dissemi-nation of that intellectual atmosphere usually roughly termed "postmodernism", a whole literature concerning the constructed character of things thought to be "natural" (such as, and perhaps first of all, ethnic and gender identity) emerged. The main goal of the proponent of social constructionism was to pull the rug from under the feet of "essentialism" and show that the solid and immutable plane so often used as a framework for conservative – if not openly discrimina-tory – policies are the result of a more or less opaque texture of narrative con-texts and power relations (see Burr (1995) for the first general introduction to the "canon" of social construction).

However, when it was time to move to the social construction of environ-mental issues (Bird (1987) is probably one of the first appearances), everything seemed more complicated. In order to defend nature, it may in fact be useful to know what it is. While, for someone fighting ethnic or gender discrimination, the constructedness of race or gender categories is usually a straightforward start-ing point, the idea that wild nature is a cultural product was inevitably bound to seem "a heretical claim to many environmentalists" (Cronon 1995, 69). While the critique of a certain cult of the wild, as that which is remote and pristine, was meant to reveal the ideological functions of the "deceptive clarity of 'human' vs. 'non-human'" and open up a "middle ground" (Cronon 1995, 85), it was received as a blow to environmental protection. For Michael E. Soulé, a biologist of conservation, nature became the object of a double siege (Soulé 1995). As if the global advancement of industrial exploitation, deforestation and pollution, the physical attack, was not enough, now a less visible and more insidious threat came from the "social siege" led by the "postmodernists" (epito-mized, according to Soulé, by Haraway (1991) and Cronon (1995)) with the reduction of the idea of nature to social and historical processes.

A decade later, more or less unintentionally, the Anthropocene hypothesis dealt a new blow to the notion of pristine, wild or, in general, independent nature. For Crutzen and Stoermer, the new awareness of the geological scale of anthropo-genic perturbations should persuade scientists and engineers of the need to "guide mankind towards global, sustainable, environmental management" (Crutzen and Stoermer 2000, 18). According to Peter Kareiva (2011), vice-president and chief scientist of The Nature Conservancy, acknowledging that human activities have altered the most basic life processes at a planetary level necessarily entails that conservancy and environmentalism must radically change their methods and objectives. For the advocates of "new-environmentalism" (Brand 2005, 2009) or "eco-modernism" (Asafu-Adjaye et al. 2015), the end of wild nature is no reason for despair. Although there is no remaining corner of the Earth where we can find an uncontaminated natural environment, nature seems to be more "resilient" than we think. See, for instance, how "around the Chernobyl nuclear facility" (and as a matter of fact you can hardly find an environment more degraded than that), "wildlife is thriving, despite the high levels of radiation" (Kareiva *et al.* 2012). An environmentalism that is effective and up to date should therefore drop its old myths – the endorsement of the assumed frailty of the balances of nature, first of all, but also the reformist mildness of sustainable development – and face the harsh reality: "One need not be a postmodernist to understand that the concept of Nature, as opposed to the chemical and physical workings of natural systems, has always been a human construction" (Kareiva *et al.* 2012). If there is no wild nature to which to return, the only option is to move forward and aim for a "Good Anthropocene". A renewed and pragmatic environmentalism should engage with the most innovative capitalist enterprises in designing new ecosystems for a flourishing economy: "Nature could be a garden" (Kareiva *et al.* 2012).

In a much-discussed book, Emma Marris developed the notion that a "global half-wild rambunctious garden" would be the model best suited to our "post-wild world". In fact, the issue is not so much to abandon the notion of wild nature, but to bring it down the pedestal, and look at it as one possibility among others:

> In different places, in different chunks, we can manage nature for different ends – for historical restoration, for species preservation, for self-willed wildness, for ecosystem services, for good and fiber and fish and flame trees and frogs.
>
> (Marris 2011, 245)

The eco-modernism of the Breakthrough Institute sparked fierce opposition from varied perspectives. Evolutionary ecologist Tim Caro (Caro *et al.* 2011) published an inventory of the regions of the world where it was still possible to find intact ecosystems that had not been not significantly altered by human activities. Eco-modernism was also pictured as a neoliberal strategy to neutralize an already ailing orthodox environmentalism (Kingsnorth 2012, 2013) and, far from disputing it, finally validated the complete domination of the Earth by human beings (Crist 2013).

Gardens and human agency in the Anthropocene: this volume

Building on this ongoing discussion, this volume aims to contribute to the interdisciplinary debate on the Anthropocene by bringing together a group of researchers from the arts, humanities and social sciences around a common case study – gardens – that acts as the starting point for reflecting on the relations between human beings and nature.

It proposes to approach gardens in a renewed way both by building on the interplay of the conceptual frameworks provided by the history of science and technology and garden and landscape studies, as well as by convening them as hermeneutic tools – not just a metaphor – to shed new light on the Anthropocene. It is also argued that a two-way reassessment is particularly fruitful, as one can simultaneously use the conceptual apparatus being developed within discussions on the Anthropocene to revisit and reassess former concepts associated with gardens and landscapes as localities where nature and culture converge and interbreed.

Gardens are used in this volume in the sense of designed landscapes of mediation between nature and culture as they embody different levels of human control over wilderness, define specific rules for this confrontation and stage different forms of human dominance. This volume's authors focus on ways of rethinking the garden and its role in contemporary society, and propose to use it as a crossover platform between nature, science and technology. Coming from different research areas – the history of science and technology, environmental studies, gardens and landscape studies, urban studies, and visual and artistic studies – they explore their joint potential, and specifically their ability to unveil various entanglements woven in the past between nature and culture, located at the antipodes of the simplistic assumptions behind the traditional nature–culture dichotomy, and they probe the potential of alternative epistemologies to get out of the predicament of fatalistic dystopias often revolving around the Anthropocene debate.

The road to this volume

As with most edited volumes, previous activities converged to the materialization of this book. From 2014 onwards, editors participated in different capacities in various Anthropocene Campi, a joint organization of the Haus der Kulturen der Welt (HKW) and the Max Planck Institute for the History of Science (MPIWG). These *campi* provided interdisciplinary *fora* for delegates of different ages, countries and areas of expertise to approach the topic of the Anthropocene in thematic and experimental seminars, whose deliverables have been feeding in part a lively webpage entitled the Anthropocene curriculum.[3]

The initial Berlin meetings (the Anthropocene Issue, 2014, and the Technosphere Issue, 2016) gave way to different worldwide offshoots, one of which took place in Philadelphia in autumn 2017. It was sponsored by the HKW and the MPIWG, and involved the collaboration of several institutions,[4] including the research unit to which all editors belong – the Interuniversity Centre for the History of Science and Technology (CIUHCT) – which brings together scholars from the two Portuguese institutions, the University of Lisbon and the NOVA University of Lisbon (both in Portugal).

Parallel to set the Anthropocene campi, in early 2016, the year prior to the Philadelphia meeting, the editors of this volume organized and/or participated in an international meeting held in Lisbon entitled "*Ars* versus *Natura*". The promise to explore the inputs from the history and philosophy of science and technology together with those stemming from gardens and landscape studies, our various fields of expertise, guided us. Participants in the "*Ars* versus *Natura*" workshop were invited to take as their starting point gardens and landscapes and/or their representations in various media as special (depicted) sites in which nature and *ars* intermingled throughout time and place. Using a canonical definition of gardens as spaces where human beings operate artificially over nature, thus opposing wilderness, the workshop encouraged participants to discuss the historical origins of such dichotomy, probe its advantages and shortcomings, its limits and potentialities, explore the interconnections between its terms, and the constraints it imposes upon thinking about gardens and landscapes, especially when it comes to shed light on issues associated with the age of the Anthropocene.

Through these two different routes – the Anthropocene *campi* and the "*Ars* versus *Natura*" workshop – the editors of this volume became convinced that by calling attention to the contingency of historical events which came to confer a hegemonic status on technoscientific epistemology, together with the concomitant necessity to historicize its associated concepts – progress, growth, *naturalization* of technology and *commodification* of nature – a whole range of possibilities opens up to historical scrutiny, including the discussion of alternative epistemologies in illuminating solutions to circumvent a no-return situation, often convoked in the debates over the Anthropocene.

The structure of this volume

Thus, in this volume, we approach the classic topic of gardens and landscapes in a novel way, by interweaving different perspectives and connecting them to the ongoing international debate on the concept of Anthropocene. In Part I. "Rethinking the garden", we revisit gardens and landscapes, real or imagined, enriched by conceptual tools drawn from the Anthropocene debate; in Part II, Gardening the Anthropocene", we take the inverse perspective and explore how the Anthropocene debate has appropriated ideas and metaphors stemming from gardens and landscape studies, and how they guide us in probing various aspects of the Anthropocene. Finally, in Part III, "Staging the Anthropocene", we extend the approach of Part II to the Anthropocene as a performative study object.

Part I approaches urban gardens as tools for enforcing technoscientific-driven worldviews, bringing to the forefront the hidden dimension of gardens as tools to enlarge the scope of human domination over nature. Far from being "sanctuaries" of nature's agency, where the natural world runs freely, the gardens presented in the four chapters comprising Part I are human-tailored landscapes designed to serve specific goals. These manicured gardens – in a broad sense, as they include not only the obvious tidy and neat flower beds and lawn arrangements, but also artificially enhanced edible plants as well as controlled wilderness – obey the various rules and constraints, taming nature and forcing it to disclose its latent power. They are examples of human self-reliance embodying the essence of the concept of *Technik*, or in Heidegger's words, a refusal "to let earth be an earth" (Vinegar and Boetzkes 2014, 143).

In Chapter 1 – "Hygiene, education and art: Roberto Burle Marx's 1930s early modern gardens in Brazil" – Aline de Figueirôa Silva approaches the early public gardens designed by the Brazilian landscape architect Roberto Burle Marx in the 1930s, in the city of Recife (Brazil). Known for his outdoor landscapes that extended the concept of the architecture of buildings into gardens and other public spaces, Burle Marx designed green spaces as places to be experienced by humans from both an aesthetical and sensorial perspective, creating scenarios where tensions and drama played an important role. His themed gardens included both plants and animals and privileged indigenous vegetation species – he was one of the first to speak out against deforestation – better adapted to the Brazilian climate. The challenge which Burle Marx's gardens present to the reader and the reason behind their particular suitability to shed light on the joint articulation of gardens and the Anthropocene lies in their unique and challenging combination of tamed nature, non-organic and synthetic materials, and ecological concerns. In a way – and appropriating Oswaldo de Andrade's concept of anthropophagy (Andrade 1928) – Burle Marx "cannibalized" the human industrial footprint (skyscrapers, huge concrete-made metropolises and paved roads along the coasts) and, after distilling it, re-enacted it in his provocative gardens.

In Chapter 2 – "Between the nuclear lab and the backyard: artificially enhanced plant breeding and the British Atomic Gardening Movement" – Vanessa Cirkel-Bartelt explores the garden as a scientific laboratory in which both scientists and gardeners participated in evaluating the results of radioactive breeding. These technoscientific gardens, which bring together formal scientific expertise and the more hands-on knowledge of gardening, are discussed in the more global framework of technofixes and public opinion. Muriel Howorth and her British Atomic Gardening Society present themselves as a solution to the problem of food crises, which, in turn, resulted from the heavy and unbridled exploitation of worldwide resources. As in all technological fixes, radioactive breeding, and, later on, atomic gardens, were adaptations, temporary solutions to a problem, and as such did not tackle its true core. Although Howorth's proposals presage what became one of the more controversial and criticized technofixes – genetically modified organisms (GMOs) – they did not receive any strong criticism at the time. Most probably the small-scale and local dimension of the project did not make visible its externalities, its negative unforeseen or unintended consequences. Nevertheless, this singular case study nicely exemplifies the hubristic belief, which lies at the heart of the debate on the Anthropocene, according to which human beings can always fully control nature.

In Chapter 3 – "Urban utopias and the Anthropocene" – Ana Simões and Maria Paula Diogo discuss urban utopias put forward in the early twentieth century, contrasting Portuguese techno-scientific and British environmental perspectives. The traditional approach that conveys a dichotomous perspective between environment-driven and industry-driven utopias is replaced by a new interpretation grounded on the concept of Anthropocene. What is at stake in this chapter is not the discussion of the usual ideological differences between

Saint-Simonism and Georgism or between industry and land, but the call of attention to their shared vision of nature as a central tool for solving problems, mostly in the realm of health and hygiene. The "engineered cities" of Melo de Matos, Fialho de Almeida and Ebenezer Howard use gardens and green landscapes not per se, but as part of a wide concept of sanitation infrastructures that counteracted the unhealthy life in overcrowded cities. They acted as buffer zones to absorb the smoke and other airborne pollutants created by industry, keeping workers healthy. The main concern of Matos, Almeida and Howard was people, not nature; nature's importance was subsumed under human agency.

This rationale extends to Chapter 4 – "Shaping colonial landscapes in the early twentieth century: urban planning and health policies in Lourenço Marques" – by Ana Cristina Roque. European gardens in African colonial settings acutely reveal how anthropogenic landscapes convey specific ideologies and assert human dominance over natural and human landscapes. The greening of colonial territories is a particularly rich topic to discuss dominance as a multi-layered concept and to take a fresh look at the contributions of social ecology to the Anthropocene debate. In Lourenço Marques (now Maputo, Mozambique), taming nature by draining swamps and imposing a disciplined European order in gardens was, in fact, only part of a larger system of domination grounded on economic and political goals. As other large technological systems – railways, roads, harbours – urban infrastructures of sanitation (gardens included) were tools for managing colonial territories in a growing global economy. Within this framework nature became a resource to assist social domination. The gardens in Lourenço Marques were part of a hygiene policy that aimed to both keep mosquito populations low (to combat malaria) and hide the native populations away in the suburbs, shaping a segregated "white city" that embodied the idea of progress and civilization advocated by the European colonial system.

Part II – "Gardening the Anthropocene" – approaches the garden as a metaphor to Planet Earth, proposing a renewed insight into the influence of the Anthropocene on the planet and on how gardeners' attitudes and expertise can offer effective alternative solutions to it. It presents on the one hand a more theoretically oriented approach to the concept of the garden, moving from the idea of an artificialized piece of nature around a house or in an urban plan to a planetary scale, and on the other hand raising a set of questions around the moral and professional scope of the gardener's work. From the idealistic visions of nature free from human agency to the real projects of reversing desert-like regions, "Gardening the Anthropocene" brings the gardener's rationale to the question of how humans should use the planet, offering a cutting-edge perspective on the co-construction of nature and humankind and its impact on future generations. In a nutshell, "Gardening the Anthropocene" points to new paths to overcome the human/nature dichotomy by entangling ecosystems and social formations.

In Chapter 5 – "From *Pairidaeza* to Planet Garden: the *homo-gardinus* against desertification" – Ana Duarte Rodrigues approaches the garden through the conceptual framework of *Pairidaeza*, the ancient Persian garden, taken to be an

---

enclosed space separated from wild nature and artificialized through human agency. She argues that the garden metaphor gained momentum as Planet Earth became threatened by the unrestrained use of its resources, a menace epitomized and popularized by the image of the Earth seen from the Moon as a fragile and finite planet. This chapter champions the philosophy of the gardener and his or her lifestyle against the *homo economicus* who tries to foster steady use and profit stemming from (unrestrained) consumption and production. The topic is addressed through several horticultural projects developed to fight desertification. The patient work of the gardener who planted trees as illustrated in the 1953 fictional story by Jean Giono, *L'homme qui plantait des arbres* [*The Man Who Planted Trees*], offers guidelines for analysing human beings' ability to change the landscape and its impact on social transformations. Thus, in this chapter, the gardener's attitude on nature abides by the good Anthropocene, and guides the discussion of real cases that evoke Giono's tale, namely the one of men who planted trees and operated an environmental and social transformation in the very precise localities of Burkina-Faso, or at a continental scale, the construction of tree curtains along thousands of kilometres to stop the process of desertification in North America, China and Africa.

In Chapter 6 – "From *Homo faber* to *Homo hortensis*: gardening techniques in the Anthropocene" – Astrid Schwarz focuses on the potential of gardens, garden practices and gardeners' values to overcome planetary destruction in the Age of the Anthropocene. The gardener is presented as an actor capable of experimenting with nature, thus embodying a specific type of *Homo faber* whose expertise is required to deal with the negative effects of the Anthropocene. *Homo hortensis* (the gardener), just like *Homo faber*, answers the "how" questions, thus fostering efficient and practical solutions to real-life problems. Gardeners have always privileged nature and space management, but recent gardening trends opt for making the best of their spatial settings, cooperating with nature, and respecting its cycles and rhythms. Therefore, *Homo hortensis'* attitude is not only embedded in values of care, esteem and survey, but also of adequacy and sustainability. Moreover, the author highlights the expertise behind various gardening practices, such as seedling, grafting, fertilizing, irrigating, as a source of inspiration for geo-engineering to match UNESCO's goals. In addition, gardening embodies a certain kind of morality, of perseverance and of values of confidence and hope for the future – subsumed in the wisdom behind the act of planting and the ability to wait for (uncertain) results that are decisive – to be confident about the possibilities offered by the uncertainties of the future. Finally, this chapter discusses how gardening can be a metaphor for "laboratory life", envisioning gardeners as the "new scientists" who control the co-construction of nature and human beings so as to secure a better life on the planet.

In Chapter 7 – "The distant gardener: remote sensing of the planetary potager" – a metaphor similar to the "Planet Garden" is used by Nina Wormbs and Johan Gärdebo. By opting for "planetary potager" rather than "Planet Garden", the authors emphasize the central role of the land as the source of humanity's livelihood and argue that Planet Earth is to be taken care of by its inhabitants as their vegetable garden. As such, they emphasize how the high technology of remote sensing was used to map the resources of the planetary potager. However, this research proved that fieldwork is indispensable, especially when the environment is more heterogeneous than expected. In view of this, the local became essential to solving global problems and once more the garden's scale becomes central to the authors' goals. From local to global and back to local, human agency on the Earth at a global scale is envisioned through spatial technology but also enables one to tackle horticultural techniques used *in loco* to inform and reframe planetary technologies. As such, this chapter provides insights into the capacity of science and technology to deal with surveys of planetary resources as well as with environmental fragility by adopting gardening and the gardener's attitudes and expertise.

Finally, in Chapter 8 – "Resistance in the garden: nature and society in the Anthropocene", Davide Scarso offers an overview of the Anthropocene's state-of-the art bibliography and compares two case studies addressing ways to surpass the classical dichotomy between nature and culture. Taking as a starting point for his reflection the utopic project of the "sustainable Half-Earth system" proposed by the American biologist Edward O. Wilson, Scarso criticizes the dichotomy of human beings from nature as impractical. Moreover, the nature–culture dichotomy neither fits nor provides answers to the dark side of the Anthropocene, since the dominance of human agency over the planet implies an entanglement of nature and humanity's imprint. Scarso also contends that the Marxist dialectical materialism of the Swedish historian Andreas Malm in *The Progress of this Storm* does not offer better or more practical solutions than those proposed by Wilson. Focusing on the segregation of nature and society, Malm argues that their combination and the concomitant blurring of their boundaries is a disaster for humankind. Although recent data on global ecosystems urge humanity to find new models of development, both Wilson's and Malm's proposals based on the separation of two poles, whether nature–culture or nature–society, do not provide answers to the planet's problems, since nature, culture and society are irremediably entangled components of Planet Earth.

In view of this, the metaphor of the garden offers a plausible terrain for a renewed conceptualization due to the garden's potential, above all other instances, as the *locus* where nature and culture meet and become integrated, and the possibilities offered by extending the garden's metaphor to Planet Earth. Building on Scarso's argument, one may stress that a garden is not, however, a hybrid, since all its elements keep their own identities: following the sixteenth-century author Bartolomeu Taegio, a garden "is the third nature" (Hunt 2000, 32), the epitome of artificialized nature, following agriculture, taken to be the second nature, and *natura natura*, the wild and first nature.

The chapters included in Part III – "Staging the Anthropocene" – explore the intertwining of Anthropocene studies, gardens and various artistic practices. Far from providing mere illustrations or parallels, the arts, and in general the question of aesthetic representation, are particularly relevant to the issues that emerge in the Anthropocene debate, since these are often permeated by phenomena which, for their complexity and their scale, are inherently "imperceptible". It is not fortuitous, therefore, that all four chapters in Part III are the result of a collaboration between two or more authors with different backgrounds or acting in more than one capacity: curator, environmental historian, artist, ecocriticist, STS scholar.

Chapter 9 – "A new machine in the garden? Staging technospheres in the Anthropocene" – by Nina Möllers and Luke Keogh, curators and historians of technology and of the environment, respectively, and historian of science and technology Helmuth Trischler, travels through several highlights of the exhibition *Welcome to the Anthropocene. The Earth in Our Hands*, held at Deutsches Museum (Munich) from December 2014 to September 2016. The authors, involved at different degrees in the conceptualization, design and management of the exhibition, show how the very act of staging the Anthropocene reveals profound parallels with gardening, since it is an activity that involves many different but interdependent processes whose results are not known beforehand. Recapitulating the most relevant concepts that structured the exhibition, this chapter singles out several key issues of the debate on the Anthropocene. Representing the complex question of the beginning of the Anthropocene, for instance, constituted a provocative challenge for a public exhibition. Far from mere technical matters, the different proposals stand on subtle and intricate connections between technological innovation and social transformation. Particular attention was given to portraying the emergence of a bio-geo-technosphere of interrelated processes as steps in an evolutionary process, the pace of which greatly increased during industrialization. An early twentieth-century Wardian case, for instance, offers a striking example of a turning point in the globalization of vegetal species, in which an apparently simple innovation actually altered the mobility networks on a global scale for centuries to come. Defining and representing "the age of humans", however, is not only a matter of what brought us here, of course, but also of what comes next, and this is where artistic experimentation is an inescapable source of insight. The exhibition included works by the Next Nature Network collective, a provocative display of disposable razor-blades as specimens of a "natural history" set-up, and the alluring "Anthropocenic Specimen Cabinet" of Yesenia Thibault-Picazo, a speculative exploration of the far future through a collection of imagined sediments. As another token of the puzzling convergence of human and geological time scales, *Welcome to the Anthropocene* also included a model of the "Clock of the Long Now", the ongoing project of designing and building a clock that may keep time for 10,000 years or more. The demanding technical and social implications of such an endeavour offer a precious and remarkable opportunity to discuss the unprecedented scale of the challenges we face. As the authors argue, fully acknowledging an inescapable quota of contingency and unruliness allows for an approach to the Anthropocene as a chance to decentre the Anthropos, rather than indulging in savagely optimistic technological utopias.

Chapter 10 – "The atom in the garden and the apocalyptic fungi: a tale on a global nuclearscape (with artworks and bird-songs)" – by science and technology studies scholar Jaume Valentines-Álvarez and figurative artist Eric LoPresti possesses a distinct performative character, pushing the formal boundaries of academic literature to their limits. Calling for a multimodal experience, the chapter proposes an alluring interweaving of critical history scholarship on nuclear energy and anti-nuke movements with paintings of the *Lewisia rediviva* flowers that today bloom in the barren soils of the Nevada Atomic Test Site, a plant bringing survival and resistance through its very name. Moreover, each section of the text is accompanied by the suggestion of a corresponding musical score. Mushrooms, also particularly resilient to radioactivity, haunt the whole text and are evocated through many different angles, as image, metaphor, food, symbol and psychotropic substance. The authors juxtapose the "global techno-Eden" envisioned at the beginning of the atomic age with the nuclear Eden of "technocratic sustainability" proposed by the advocates of a Good Anthropocene. Conjuring up a vivid imagery that cuts across society, nature and technology, they contrast the giant deadly mushroom of nuclear detonation with the sparse underground networks of fungal mycelia, bringing to the fore the meshwork of many mutual aid-based social experiences that blossom in the harsh environments created by natural and human-induced catastrophe all over the globe.

In Chapter 11 – "Inhabitants: image politics in ongoing climate crisis", artists and cultural activists Mariana Silva and Pedro Neves Marques present the online video channel they founded in 2015 and which distributes through several online services short videos merging contemporary art with research journalism. In A *Brief History of Geoengineering*, for instance, in order to bring to the fore the interrelation of economic interests and marketing strategies that underpin a seemingly "technical" issue, several layers of alteration follow the speaker's argument and disturb a flow of patent applications images with background noises and CGI weather effects. The series on the struggles concerning oil extraction in Portugal, *For an Oil Free Future*, adopted a different approach, and, by mixing talking-head format with dystopian fiction, it aimed at both intervening in the debate and providing material for militant action. The current series *What is Deep Sea Mining?*, co-authored with art curator Margarida Mendes, introduce the audience to the emergent possibility of extracting valuable metals from certain mineral formations on the seabed. Explanations, case studies and interventions from experts and activists portray the many technical and environmental issues at stake. Furthermore, the fact that deep ocean habitats are still largely unexplored and that deep sea mining has yet to actually take place constitute a formidable challenge to image making. At the threshold that separates and joins research, experimentation and activism, the over 25 videos produced thus far not only constitute a bold intervention in several crucial environmental struggles, but also, and at the same time, a sophisticated experimental reflection on visibility, representation and the very nature of the video form in an "imagetic-ideological regime" dominated by market-driven virality.

In Chapter 12 – "Troubled gardens: nature–technoculture binary and the search for a Safe Operating Space in Hayao Miyazaki's *Mononoke Hime*" – by environmental engineer Ivo Louro and artist and essayist Ana Matilde Sousa, both acting at the crossroads of environmental studies, cultural theory and artistic practice, explores the representation of gardens in the Anthropocene by analysing the animated movie *Princess Mononoke*, written and directed by Japanese author Hayao Miyazaki, co-founder of the celebrated Studio Ghibli. Louro and Sousa focus in particular on two specific locations of the film: the secret garden of Lady Eboshi in her Iron Town and the equally secluded pond of the elk-like Forest Spirit in the depths of the Cedar Forest. The two "gardens", as the authors convincingly argue, embody the essential conflict of opposite forces that structures the whole narrative: wilderness and civilization, human society and nature, growth and equilibrium. Moreover, the chapter shows how, in *Princess Mononoke*, gardens function as enclosed spaces of difference that, in their reciprocal connections, embody the inescapable interrelation of natural and human agencies. Unlike many contemporary representations of the conflict between nature and technological society, Myazaki's movie manages to avoid both a romantic nostalgia of lost harmony and eco-modernist optimism. Louro and Sousa show how in *Princess Mononoke* the intricate entanglements of nature, culture, gender and race all concur in portraying the acknowledgement of difference as essential to life itself. Based on this insight, the authors then offer a valuable qualification of the notion of "Safe Operating Space", stressing its inherently tense precariousness.

To sum up, all chapters in this volume encourage readers – from either an academic background or a larger audience – to engage in the debate on nature and human agency. Using specific case studies centred on the concept of garden, an "object" that we all know and experiment with, the authors propose to overcome simplistic dichotomies that identify nature with good and human actions with evil, and question the surreptitious a-historicism and determinism tone that often "colours" the debate and the discourses on the Anthropocene. The chapters in this volume use gardens and gardeners to analyse environmental issues as politicized phenomena driven by human agency and shaped both by individual and collective economic, political and cultural decisions. The garden is a negotiated space between humans and nature, and as all negotiations tell us, their outcome is not necessarily a win-win situation.

Notes

1 Not only is knowledge production indelibly tied to specific sites or venues, but there are also connections between more or less extended regions (provinces, countries) and the emergence of specific technoscientific cultures. Furthermore, the movements and circulation of technoscientific knowledge also put in evidence how encounters give rise to negotiations occurring in specific sites, accounting for instances of appropriation and transformation of knowledge, for the development of techniques of disciplining the senses, of securing the credibility and authority of travellers, scholars and mediators, and of reproducing and representing the unfamiliar and the novel.
2 See www.mpiwg-berlin.mpg.de/research/projects/DEPT1_Valleriani_Pratolino.
3 See www.anthropocene-curriculum.org/ for details.
4 Drexel University, the Society for the History of Technology, the Chemical Heritage Foundation, the Technische Universiteit Eindhoven, Virginia Tech, KTH Royal Institute of Technology.

Bibliography

Andrade, Oswaldo de. 1928. "Manifesto Antropofágico". *Revista de Antropofagia*, 1(1).

Asafu-Adjaye, John, Linus Blomqvist, Stewart Brand, *et al*. 2015. *An Ecomodernist Manifesto* (online). Oakland: Breakthrough Institute. Retrieved from www.ecomodernism. org/manifesto-english/.

Baldassarri, Fabrizio and Oana Matei. 2017. Gardens as Laboratories. The History of Botany through the History of Gardens (Special Issue). *Journal of Early Modern Studies*, 6(1).

Benjamin, Walter. 1986. *Reflections: Essays, Aphorisms, Autobiographical Writings*, edited by Peter Demetz. New York: Shocken.

Bird, Elizabeth A.R. 1987. "The Social Construction of Nature: Theoretical Approaches to the History of Environmental Problems". *Environmental Review*, 11(4): 255–264.

Bloch, Marc. 1931. *Les caractères originaux de l'histoire rurale française*. Paris: Belles Lettres.

Bookchin, Murray. 1982. *The Ecology of Freedom: The Emergence and Dissolution of Hierarchy*. Palo Alto, CA: Cheshire Books.

Brand, Stewart. 2005. "Environmental Heresies". *MIT Technology Review* (online). Retrieved from www.technologyreview.com/featuredstory/404000/environmental-heresies.

Brand, Stewart. 2009. *Whole Earth Discipline: An Ecopragmatist Manifesto*. New York: Viking.

Braudel, Fernand. 1949. *La Méditerranée et le Monde Méditerranéen a l'époque de Philippe II*, 3 vols. Paris: Armand Colin.

Burr, Vivian. 1995. *An Introduction to Social Constructionism*. London: Routledge.

Caro, Tim, Jack Darwin, Tavis Forrester, Cynthia Ledoux-Bloom and Caitlin Wells. 2011. "Conservation in the Anthropocene". *Conservation Biology*, 26(1): 185–188.

Crist, Eileen. 2013. "On the Poverty of Our Nomenclature". *Environmental Humanities*, 3: 129–147

Cronon, Willliam, ed. 1995. *Uncommon Ground: Rethinking the Human Place in Nature*. New York: W.W. Norton.

Crutzen, Paul and Eugene Stoermer. 2000. "The "Anthropocene". *Global Change Newsletter*, 41: 17.

Febvre, Lucien. 1935. "Pour l'histoire des sciences et des techniques". *Annales d'histoire économique et sociales*, 7(36): 646–648.

Hacking, Ian. 1999. *The Social Construction of What?* Cambridge, MA.: Harvard University Press.

Haraway, Donna J. 1991. *Simians, Cyborgs, and Women: The Reinvention of Nature*. New York: Routledge.

Hillman, Mayer. 2018. "'We're Doomed': Mayer Hillman on the Climate Reality No One Else Will Dare Mention". *Guardian*, 26 April. Retrieved from https://..//2018/apr// were-doomed-mayer-on-the-climate-reality-no-one-else.

Hunt, John Dixon. 2000. *Greater Perfections: The Practice of Garden Theory*. Philadelphia: University of Pennsylvania Press.

Kareiva, Peter. 2011. "Failed Metaphors and a New Environmentalism for the 21st Century". National Academy of Sciences, Beckman Center – Irvine. Retrieved from www.youtube./?feature=player_embedded&v=4BOEQkvCook.

Kareiva, Peter, Michelle Marvier and Robert Lalasz. 2012. "Conservation in the Anthropocene: Beyond Solitude and Fragility". *Breakthrough Journal*, 2 (online). Retrieved from https://.org/.php/journal/past-issues/issue-2/conservation-in-the-anthropocene.

Kennedy, Tara. 2014. "Heidegger and the Ethics of the Earth: Eco-phenomenology in the Age of Technology". Retrieved from http://digitalrepository.unm.edu/phil_etds/7.

Kingsnorth, Paul. 2013. "Dark Ecology". *Orion Magazine*. Retrieved from https://orion-magazine.org/article/dark-ecology/.

Kranzberg, Melvin and C. Pursell, eds. 1967. *Technology in Western Civilization*, 2 vols. New York: Oxford University Press.

Livingstone, David N. 2003. *Putting Science in Its Place. Geographies of Scientific Knowledge*. Chicago, IL: University of Chicago Press.

Livingstone, David N. 2010. "Landscapes of Knowledge". In *Geographies of Science*, edited by Peter Meusburger, David Livingstone and Heike Jöns (pp. 3–22). Dordrecht: Springer.

Malm, Andreas. 2018. *The Progress of this Storm. Nature and Society in a Warming World*. London: Verso.

Marris, Emma. 2011. *Rambunctious Gardens. Saving Nature in a Post-wild World*. New York: Bloomsbury.

McKibben, Bill. 1989. *The End of Nature*. New York: Random House.

Picon, Antoine. 2010. "Nature et ingénierie: le parc des Buttes-Chaumont". *Romantisme*, 150(4): 35–49.

Pollan, Michael. 1991/2003. *Second Nature. A Gardener's Education*. New York: Grove Press.

Proctor, James D. 1998. "The Social Construction of Nature: Relativist Accusations, Pragmatist and Critical Realist Responses". *Annals of the Association of American Geographers*, 88(3): 352–376.

Schatzki, Theodore. 2003. "Nature and Technology in History". *History and Theory*, 42(4): 82–93.

Scranton, Roy. 2018. *We're Doomed. Now What? Essays on War and Climate Change*. New York: Soho Press.

Soulé, Michael E. 1995. "The Social Siege of Nature". In *Reinventing Nature? Responses to Postmodern Deconstruction*, edited by Michael E. Soulé and Gary Lease (pp. 137–160). Washington, DC: Island Press.

Vinegar, Aron and Amanda.Boetzkes, 2014. *Heidegger and the Work of Art History*. Aldershot, Surrey: Ashgate.

Williams, James. 2010. "Understanding the Place of Humans in Nature". In *The Illusory Boundary: Environment and Technology in History*, edited by Martin Reuss and Stephen H. Cutcliffe (pp. 9–25). Virginia: University of Virginia Press.

Williams, Rosalind. 2012. "The Rolling Apocalypse of Contemporary History". In *Aftermath: The Cultures of the Economic Crisis*, edited by M. Castells, J.M.G. Caraca and Gustavo Cardoso (pp. 1–24). Oxford: Oxford University Press.

Part I
Rethinking the garden

Establishing the scene

1 Hygiene, education and art

Roberto Burle Marx's 1930s modern gardens in Brazil

Aline de Figueirôa Silva

Introduction

The Brazilian artist Roberto Burle Marx (1909–1994) is considered to be one of the leading landscapers of the twentieth century. His vast work ranges from squares to large-scale parks, public and private gardens, paintings, tapestries, crystals, jewels and other artefacts. During a period of intense cultural debate, renewal of the arts and construction of Brazil's national identity, Burle Marx conceived, based on principles of painting, botany and ecology, his concept of the modern garden by using native vegetation as a means for enhancing Brazilian roots.

His main principles were already in place when he designed his first public gardens (Recife in the 1930s), which according to Burle Marx himself "was fundamental to the course that my professional activity took on" (Marx 1985, 70). In addition, and in what concerns the use of native plants, one of his main sources of inspiration was the work of Auguste François Marie Glaziou, the French landscaper at the service of the Portuguese Royal Court in Brazil, author of the main nineteenth-century gardens in Rio de Janeiro, and "precursor of the botanical travels that explored the inland of Brazil" (Sá Carneiro 2017, 82).

Roberto Burle Marx was born in São Paulo in 1909. He was the son of Cecília Burle, a member of a traditional family of French ancestry, and Wilhelm Marx, a German Jew from Stuttgart. The family moved to Rio de Janeiro in 1913. Even as a child, he nurtured an uncommon interest in plants and came to develop his taste of botany by observing the species in the garden of his parents' home and by reading the German journal *Gartenschönheit*, brought from Europe by his father, and which allowed him to get in touch with parks and gardens in other countries, as well as with Brazilian plants that revealed to him "a world hardly known" (Marx 1985, 71).

In 1928, Burle Marx travelled to Germany for medical treatment for eye problems. In Europe, he went to concerts, visited exhibitions, and attended music and painting classes. In Berlin, he visited the greenhouses of the Botanical Garden of Dahlem, where he first became acquainted with the Brazilian native flora, gazing at its incredible potential. It was then, "before a greenhouse of Brazilian tropical plants", that he took "the decision to build, with the native flora, a whole new order of plastic composition, for drawing, for painting, and even reaching the landscape and the garden" (Marx 1954, 18). Upon returning to Brazil in 1930, he attended a course on painting at the *Escola Nacional de Belas Artes* (National School of Fine Arts) in Rio, directed by the architect Lucio Costa. Two years later, he designed his first garden for a private residence – the Schwartz house – by the architects Lucio Costa and Gregori Warchavchik.

However, it was in the city of Recife, the capital of the northeastern Brazilian state of Pernambuco, that he designed and created his first public gardens.[1] Appointed as Head of the Department of Parks and Gardens of the Architecture and Building Department (DAC) of the Government of Pernambuco upon Lucio Costa's suggestion, Burle Marx was in charge of designing and remodelling several public gardens in Recife. In total, between 1935 and 1990, he submitted 58 projects in Pernambuco, at least 21 of which – 17 public and 4 private gardens – dated from the period between 1935 and 1937 (Sá Carneiro, Silva and Silva 2013, 243–245).[2] Out of the 17 public gardens, 8 were in fact built or remodelled, 3 never came to life, and it is unclear what happened to the remaining 6 (Sá Carneiro, Silva and Silva 2013, 243–245). Burle Marx's plan for public gardens in Pernambuco included 15 squares, a zoological and botanical garden, and a public park. These new public gardens – both new projects in suburban neighbourhoods and remodelled pre-existing gardens located in the historic centre of the city – used open spaces already laid out in the urban fabric.

During his time in Pernambuco, Burle Marx grew close to other young artists and intellectuals such as the architect Luiz Nunes, the engineers Antônio Bezerra Baltar, Attilio Corrêa Lima, Ayrton de Carvalho and Joaquim Cardozo (also a poet), the writer Clarival do Prado Valladares, and the sociologist Gilberto Freyre. In Recife, he wrote articles, gave interviews, attended parties and got in touch with local traditions. The landscape of Recife and the profile and functions of gardens in human history as a sign of human agency were the main topics of his articles. In *"Jardins para o Recife"* ("Gardens for Recife"), published in the *Boletim de Engenharia* (Engineering Bulletin) in March 1935, the landscaper stated:[3]

> A garden is in its essence organized nature, subordinated to architectural laws. [...] Gardens across all times and among all peoples, emerged at the omega of their respective civilizations. [...] All relevant cities have gardens. Thus, one may conclude that gardens are rather a conscious necessity than simply an accidental creation of superfluous luxury in our civilization. In every garden there is always a plan for ordering nature. What changes across time is just its spirit. [...] The modern garden does not escape this logic. This is why it encompasses several objectives: hygiene, education and art.
>
> (Marx 1935a, n.p.)

As such, Burle Marx envisaged gardens as architectural creations, emphasizing the historical dimension of their uses and stressing that modern gardens were part of this long tradition by fulfilling specific functions related to hygiene, education and art.

Hygiene, education and art: Roberto Burle Marx's modern garden foundations

By anchoring his understanding of the modern garden on the trio hygiene-education-art, Burle Marx emphasized vegetation as one of its main protagonists. Plants were the main actors both for conveying the artistic and cultural messages underlying the landscape arrangement and for structuring outdoor urban leisure spaces in a tropical country such as Brazil, and particularly in a hot and humid city such as Recife.

Hygiene-wise, the modern garden stood for "a true collective lung" for the working class, as well as for the less well-off children. It was a space where "the urban inhabitant came to breathe a little fresh air, tired of the daily struggle in small offices, paved streets and factory environments", and where "children living in perched apartments, small yard houses or collective dwellings" could enjoy their toys and breathe "an air free of contamination". To achieve these goals "in tropical climates" it was "indispensable" to introduce into gardens "trees able to cast huge shadows" (Marx 1935a, n.p.). From an educational point of view gardens were tools for botanical instruction, as well as spaces for nourishing a balanced relationship between city dwellers and nature. In Burle Marx's words, they should awake in the "inhabitant of the city a little love for nature and provide him with the means to distinguish local plants from exotic flora" (Marx 1935a, n.p.). Finally, concerning art, the modern garden "should comply with a basic idea and all parts should be subordinated to a coherent ensemble [...]. This harmony of the whole garden is deemed vital to human well-being" (Marx 1935a, n.p.).

Burle Marx reasserted these three principles in his 1935 article "*Jardins e Parques do Recife*" ("Gardens and Parks of Recife"), published in the newspaper *Diario da Tarde*, stating that an "integral part" of the city gardens "should fulfill a function, or to be more precise, many functions: hygienic, educational and artistic. When a garden achieves these goals, it showcases the level of culture of a people" (Marx 1935b, 1). In interpreting these three propositions, Sá Carneiro (2017, 84) synthesized Burle Marx's gardens as "a human intervention in nature, which manipulates the existing living elements, such as vegetation, water, and soil, as well as the few built elements".

Bearing in mind the trio hygiene, education and art, two premises underlie the role of plants as main actors in Burle Marx's modern garden. The first refers to plants as part of a whole; that is, their mutual relationships and the principles that rule their selection and organization *in situ*. The second concerns plants and their relationships with human beings; that is, the way users appropriate the garden.

According to the first premise, gardens should obey the laws of composition of architecture and painting – relationships of symmetry, contrasts between light and shadow, interaction among the different volumes of trees, shrubs and herbaceous plants, textures formed by canopies, trunks, foliage and flowers – and of botany and ecology – the observation of plants in their natural habitats, associations of species, compatibility between vegetation and environment, and its adaptation conditions. According to the second premise, gardens should respond to the recreational needs of urban populations and to reduce heat in tropical cities. Burle Marx conceived of gardens as privileged public spaces where city dwellers could find shelter and fresh spots to escape the hot tropical climate and, at the same time, become knowledgeable about the native flora of several Brazilian regions and, in particular, of the region where they lived.

Local landscapes and their physical, biotic and anthropic attributes (rivers, soil, relief, climate, fauna and flora, adjacent buildings, customs and cultural practices, etc.) are at the core of both premises: vegetation as an essential attribute both for the composition of gardens and their uses. These ideas were already present in the first two gardens designed in Recife in 1935: the *Praça de Casa Forte* (Casa Forte Square) and the *Praça Euclides da Cunha* (Euclides da Cunha Square).

New garden projects: the enhancement of the indigenous flora of Brazil

A *Praça de Casa Forte* (Casa Forte Square)

The Casa Forte Square was Burle Marx's first project for a public garden. The garden was designed to occupy a space known as *Campina da Casa Forte* (*campina* is an area with no trees), delimited by an area of residential buildings constructed following the deactivation of an old sugar mill. The so-called *Engenho da Casa Forte* (Casa Forte Sugar Mill) was built in the mid-sixteenth century, and as early as 1645 the whole area was considered to be one of the best agricultural properties in Pernambuco (Costa 2001, 61). The Battle of Casa Forte took place at the Campina da Casa Forte during the same year, culminating in the expulsion of the Dutch from Pernambuco following the Portuguese colonization of Brazil. Later on, the plantation heirs handed over the area to the municipality to embellish the local church, to serve as a space for commercial fairs, and to perpetuate the memory of the Brazilian victory against the Dutch (Costa 2001, 65). In the early twentieth century, the place was used for local religious events and profane celebrations. It received a memorial plaque in 1918 and a monument in the 1930s, both upon the recommendation of the journalist and cultural activist Mario Melo (Melo 1935, 6, Silva 2010, 142) in order to pay tribute to those who lost their lives on the battlefield.

In the early 1930s, the *Campina da Casa Forte* comprised a narrow and long terrain, bordered by residential buildings, with a church in one of the extremities, and was already divided into three parts, as seen in the 1932 *Planta da Cidade do Recife e Arredores* (Plan of the City of Recife and its Surroundings). During the administration of the mayor Antônio de Góis (1931–1934), the

Figure 1.1 Praça de Casa Forte (Casa Forte Square): monument erected in the 1930s to pay tribute to Brazilian soldiers killed on the battlefield in 1645, the church, flower beds, benches and palm trees.

Source: *Annuario de Pernambuco para 1935*, published in Silva (2010, 143).

area received the aforementioned commemorative monument in art deco style, art nouveau cement banks, and some garden works which included the plantation of at least two species of palm trees, coconut (*Cocus nucifera*) and Chinese palm tree (*Livistona chinensis*) (Silva 2017, 129).

In 1935, Burle Marx began to remodel the *Campina da Casa Forte* space. The Casa Forte Square, also called Casa Forte Park, was inspired, as the landscaper often mentioned, by the Botanical Garden of Berlin, photographs of Kew Gardens in London and by the *Parque de Dois Irmãos* (Park of Two Brothers) in Recife (Marx 1935c, 12, Marx 1985, 71, Hamerman 1995, 167). He conceived of three lakes, one in each section of the square, "subordinated, however, to a coherent ensemble" (Marx 1935c, 12). Each of the lakes per se performed a specific "educational function", since they represented "an isolated group, based on the geographical origin of their elements", as explained by Marx in his article "*O Jardim da Casa Forte*" ("The Casa Forte Garden") published in *Diario da Manhã* (Marx 1935c, 12).

In the same text, Burle Marx mentioned that water – another of the main elements of his project – was an overarching element in the history of gardens and had never ceased to be used in Spain, France, Germany and Italy, adding that it was André Le Nôtre, the famous French landscaper, principal gardener of King Louis XIV and author of the Versailles Gardens, who had created for the first time large water surfaces which replaced "the *parterre de broderie* by the *parterre-d'eau* with the purpose of creating calm areas, new light and reflection perspectives and contrasts"(Marx 1935c, 12). Until then, water had just been used "as an element to reflect the surrounding garden or to feed the fountains" (Marx 1935c, 12).

Having in mind this historical framework and taking into consideration the use of water "as a means capable of allowing the cultivation of an endless variety of plants", Burle Marx proposed to create a "water garden" featuring the "Victoria Regia [i.e. *Victoria amazonica*, the largest plant of the *Nymphaeaceae* family of water lilies] as the central motif" (Marx 1935c, 12). The plants used in each of the three sections and three lakes of the square were chosen according to their geographical origin. The first section of the square, next to the main avenue of the neighbourhood, was composed of native American plants, including many Brazilian species; the second stretch was entirely devoted to the Amazonian flora;[4] the third stretch, in front of the church, was the habitat of exotic plants from other continents. In the central lake, "a statue by Celso Antônio, representing an Indian bathing"[5] (Marx 1935c, 12), brought the whole garden together by establishing a bridge between human beings (in this case a native Amazonian woman) and plants belonging to the same biome. However, the sculpture was never put in place.

In the first rectangular section of the square, Burle Marx emphasized the lake as the main protagonist featuring colourful aquatic plants existing in Brazilian rivers, weirs and lakes.

> Around the lake plants such as the *aningas* [*Montrichardia linifera*] [...] the celebrated *Tajás of the Amazon* [*Caladium bicolor*, also known as elephant ears or angel wings] [...] some representatives of the grass family, etc. provide the feeling of tropical lushness. Walking from the inside out, we find a lawn and a promenade. Finally, two rows of trees such as *canafístula* [*Peltophorum dubium*, golden rain tree], *ipê* [roble], *mulungu* [*Erythrina velutina*], *munguba* [*Pachira aquatica*] etc.
>
> (Marx 1935c, 12)

The description of both the central stretch and the lake showed interest in valuing Amazonian plants and their plastic aspects, such as volume, colour and texture, especially the *capirona* (*Calycophyllum spruceanun*) planted for the first time in Recife (Cardozo 1973, 171).

> Surrounding the lake, there is a row of *capironas*, an interesting tree because of the well-defined shape of its trunks in colonnades and symmetrical canopies, of great decorative effect for architectural gardens. Besides the entrances to the promenade that surrounds the lake, one spots elephant ears [also known as angel wings] that give a note of color. In the four angles are blocks of Amazonian palms.
>
> (Marx 1935c, 12)

Figure 1.2 Praça de Casa Forte (Casa Forte Square): detail of the central lake with ama-
zonean plants. In the lake lies the Victoria Regia (i.e. *Victoria amazonica*).

Source: Fundação Joaquim Nabuco. Photo by Benício Whatley Dias, *c.*1940, published in Silva
(2010, 151).

In the last section of the square, also rectangular, Burle Marx emphasized the
use of aquatic flora from other tropical regions and the plasticity of the flowering
of exotic trees, such as flamboyants and acacias:

> The exotic lake features the aquatic flora of the tropical regions of other
> continents. One spots lotus [*Nelumbo nucifera*] [...] *Cyperus papyrus* [...].
> Among the plants one finds specimens of great beauty such as *Canna indica*,
> [...] *Crinum powellii*, strelitzia [*Strelitzia* sp.] and some decorative musaceae.
> [...] torch gingers [*Etlingera elatior*] will be planted. Among the trees that
> flank the lake are: teak wood [*Tectona grandis*], flamboyants [*Delonix regia*] of
> red and yellow flowering, various acacias [*Acacia sp.*] etc.
>
> (Marx 1935c, 12)

Biologist Joelmir Marques da Silva highlighted the "significant plant diversity" (Silva 2017, 139) at the Casa Forte Square where at least 58 species were

Figure 1.3 *Praça de Casa Forte* (Casa Forte Square): detail of the lake with exotic plants. In the background is the church.

Source: Fundação Joaquim Nabuco. Photo by Benício Whatley Dias, 1938, published in Silva (2010, 151).

registered by Burle Marx: 27 tree species, 7 palm species and 24 herbaceous species.[6] In spite of the quantity and diversity of native species, Burle Marx used exotic plants, among which the *ravenala* or traveller's palm (*Ravenala madagascariensis*) and the flamboyant were already in use in some gardens in Recife.

Concerning spatial organization, in each section of the square one found herbaceous plants by the lakes encircled by two rows of trees, thus creating different effects according to the direction of the paths – longitudinal or transverse. In the longitudinal direction, the vegetation created geographical compositions between the main road and the church, complying with the terrain's length; in the transverse direction, the vegetation created volumetric compositions and contrasted areas of shadow at the edges and of light at the centre, as the avenues of trees demarcated the space and marked the transition between the urban exterior and the architecturally organized nature of the interior.

Burle Marx topped his landscape project with modernist polished granite benches along the avenues of trees on both square sides. The shadows cast by the trees onto the benches created suitable, enjoyable and attractive spots for watching the visual relaxing effects of the water and the aquatic plants of the lakes.

In addition to being inspired by the water gardens of Kew and the Brazilian flora observed in the greenhouses of Berlin's Botanical Garden, the design of the garden was conceived to fit perfectly the geometric shape of the square already laid out in the urban fabric and to respond to the particular characteristics of Recife's waterscape (rivers, islands and wetlands). In the city, "rivers were always of great importance", as recalled by the landscaper in his last interview (Hamerman 1995, 169).

By creating distinctive scenarios that encouraged visitors to discover the flora of different geographical regions and by highlighting the sculptural forms of plants and the aesthetic value of water features, Burle Marx was mindful of the importance of walking through the garden as a dynamic visual and sensorial experience, asserting his vision of gardens as spaces of hygiene, education and art:

> [In these gardens] one will learn about the richness of the tropics, their large trees, lush foliage and intense blooms, while experiencing the benefits of cooling shades in an educational environment, the whole elements being subordinated to a global aesthetic rationale.
>
> (Marx 1935c, 12)

Burle Marx's project was not received without criticism. Journalist Mario Melo (who, as already mentioned, led the proposals for the two monuments honouring the Battle of Casa Forte's fighters) wrote in his column *Ontem, Hoje e Amanhã* (Yesterday, Today and Tomorrow), published in a local news-paper: "the history of Amazonian landscape" caused the destruction of the "monument honoring Brazilian fighters of the Dutch war" (*Jornal Pequeno* 1935b, 1), as if "worshiping the memory of our ancestors is a thing of the past" (*Jornal Pequeno* 1935a, 1). He reiterated his attacks on Marx's projects on the grounds that they were responsible for the demolition of the "monument that paid tribute to the famous battle between the liberators of Pernambuco and the Dutch" (*Jornal Pequeno* 1935d, 1). Burle Marx justified the removal of some monuments by arguing that as sculptural pieces they were of very poor quality:

> poor sculptures are often left in squares because they are statues of great men. If one really wants to immortalize a prominent figure in bronze, let's do it with art. [...] What is necessary, therefore, is that art prevails over sentimentality.
>
> (*Diario de Pernambuco* 1937)

Concerning this particular monument, he also reiterated that it was "a sculpture of very dubious artistic value" (Hamerman 1995, 169) and not worthy of the "merits of the heroes it honored" (Marx 1962, 21).

A *Praça Euclides da Cunha* (Euclides da Cunha Square)

Still in 1935, Burle Marx designed the Garden of Benfica, renamed Euclides da Cunha Square, to honour the celebrated Brazilian author of the book *Os Sertões*,[7] exploring the many plant species of the *Caatinga*.[8] This was his second public garden project, built in a space known as *Largo do Viveiro* (Viveiro

Figure 1.4 Largo do Viveiro (Viveiro Square).
Source: Brito (1917), published in Silva (2010, 154).

Square), where an old sugar mill called *Engenho da Madalena* had been located, and which hosted a sanitary station that was part of the sewage system of Recife, inaugurated in 1915. Silva (2017, 111) mentioned that a public contest was held by the municipality in 1931 to build a garden in Viveiro Square and that the winning project included the construction of a central bandstand, a pond, a fountain and a row of 15 trees. However, there is no evidence that this project was ever brought to life.

Burle Marx was commissioned to design and build a new garden in 1935. Its plan was organized around a central bed of several species of cacti bordered by blocks of stone *(cactario)*, following the same principle used at the Casa Forte Square: herbaceous plants in the centre and two avenues of trees in the periphery, marking the transition between the exterior and the interior, and a zone of light in the centre contrasting with a shadow area on the edge.

> Surrounding the *cactario*, there will be a walkway of slabs placed apart in order to let grass to grow in the spaces between them. This path allows for a close observation of various cacti specimens. Two alleys of native trees of the *sertão*, such as *umbuzeiros* [*Spondias tuberosa*], *joazeiros* [*Ziziphus joazeiro*], [...] wrap the external part of the square.
>
> (Marx 1935b, 1)

Figure 1.5 Praça Euclides da Cunha (Euclides da Cunha Square): Caatinga plants. In the background is the *Clube Internacional* (International Club).

Source: Fundação Joaquim Nabuco, published in Silva (2010, 157).

Inside, a pathway allowed users to get closer to the cacti which grew in the lower level, and which could also be accessed by three sets of steps flanked by large cacti which acted as porticoes.

The ensemble of plants at the Euclides da Cunha Square included at least 23 species, of which 10 were tree species, 7 shrub species and 6 herbaceous species (Silva 2017, 120). These species showcased to the public an important biome from Brazil, particularly from the northeast inland, and re-created the dry land-scape of the *sertão* (backlands) in a coastal city. This is why Silva (2017, 128) singled out the "ambiguous" character of the use of plants from the Caatinga in Recife, insofar as they are native to Brazilian flora, but exotic and unknown to the inhabitants of the city.

Polished granite benches were placed "under the leafy treetops", creating shade to protect visitors from the sun and inviting them to admire the cacti

(Marx 1935b, 1). The avenues, "besides providing abundant shade, create a per-spective effect that emphasizes the light projected on the cacti" (Marx 1935b, 1). A polished granite sculpture of a "man in a loincloth", again by sculptor Celso Antônio, occupied a central place among the cacti; the man portrayed had the physical traits and expression of courage typical of "the Brazilian of the North, in perfect harmony with the well-defined forms of cacti" (Marx 1935b, 1). As in the case of the statue of the Indian woman bathing at the Casa Forte Square, no sculpture was installed at the Euclides da Cunha Square.

Once again, Burle Marx's project was criticized by journalist Mario Melo, who referred to the Euclides da Cunha Square as "the *sertanização* [to become similar to the semi-arid region of northeastern Brazil] of the mangroves of the old Madalena mill" (*Jornal Pequeno* 1935c, 4) and stressed "the madness of trying to implement in Benfica a small *sertão* [...] just by contrasting vegetation with an aquatic environment" (*Jornal Pequeno* 1936, 1). He added: "if this is a garden, it is a garden without flowers, entirely made of thorns; if it is a park, it is a park without trees" (*Jornal Pequeno* 1936, 1). Mario Melo's negative comments reveal the strangeness of having in Recife, a tropical coastal city, a garden that displayed not the usual plants but a diversity of plants typical of the *sertão*, an experience never seen before in public gardens.

The Euclides da Cunha Square reified Burle Marx's design concept of the modern garden anchored on the trio hygiene, education and art, and using native tropical vegetation as its main structural element:

> This concept is at the core of our decision of building a garden in Viveiro Square, *Madalena*, that although protecting us with big shadows, enables us to get in touch with the very curious flora of the Brazilian Northeast: the cacti. [...] We hope to have achieved our goal, that is, to offer to Pernam-buco a garden in which hygiene, art, education and culture are brought together.
>
> (Marx 1935b, 1)

Planting cacti in a public garden enhancing Recife's urban landscape showed Burle Marx's concern with "culture, education and ecology": as the harsh reality and cultural richness of the *sertão* travelled to the coastal town (Sá Carneiro 2017, 89), so did the "Brazilian soul" grow "in our parks and gardens" (Marx 1935b, 1).

Interventions in pre-existing gardens: the dialogue between old and new

A *Praça da República* (República Square)

In addition to these two major projects, Burle Marx was in charge of smaller interventions in many gardens of Recife. One of the remodelling projects took place at the República Square, "perhaps the most important work entrusted to

Burle Marx" (*Diario da Tarde* 1937). Cardozo's statement was probably due to the relevance of the square located on Antônio Vaz Island, one of Recife's most notable historic sites, a unique geographical setting dominated by the Capibaribe River, and hosting many historic bridges and buildings. Under the Dutch occupation in the mid-seventeenth century and during the rule of Count John Maurice of Nassau-Siegen, the Dutch built the Palace of Freiburg and Mauritius, or *Mauritsstad*, as their government headquarters in Brazil.

After the Dutch were expelled from Pernambuco, many of its buildings were destroyed and replaced: the Palace of Freiburg was replaced by the building of the *Erário Régio* (State Treasury) which was, in turn, demolished in 1840 and replaced in 1843 by the Palace of the Pernambuco government. From the mid-nineteenth century onwards the Teatro de Santa Isabel (Saint Isabel Theatre), the Public Library, the School of Engineering (occupied by the State Treasury) and the Lyceum of Arts and Crafts were built around the square.

The República Square comprised three parts: in the west there was the Saint Isabel Theatre, in the east the State Treasury and in the north the Palace of the Pernambuco government; all three buildings surrounded the trapezium-shaped central part of the square. In 1872, as a result of urban improvements carried out by the provincial administration of Pernambuco, the central area of the square was landscaped to provide residents "with a place for gathering and recreation

Figure 1.6 Praça da República (República Square): iron railings, bandstand, imperial palm trees and flower beds. On the left is the Saint Isabel Theatre and in the background the Palace of the Pernambuco government.

Source: Instituto do Patrimônio Histórico e Artístico Nacional, published in Silva (2010, 63).

and beautifying the city" (Silva 2018, 115). Works included a railing with four iron gates imported from England, an iron bandstand, metal sculptures imported from France and four small lakes. The iron statues of 1863/1864 depicting Graeco-Roman goddesses "were produced by the French foundry J.J. Ducel et Fils, and signed by the sculptor Eugène Louis Lequesne" (Silva 2018, 115). In addition to trees, shrubs and flower beds, the garden featured "four rows of imperial palms creating two cross-shaped alleys", which interacted visually and conceptually with the surrounding buildings (Silva 2018, 120). This species, introduced and widely used during the Empire – hence its popular name – "created a tropical atmosphere and a *sui generis* effect between the garden and the urban environment" (Silva 2018, 120).

Around 1925, the square was remodelled within a more general plan for urban expansion. Photographs from the mid-1920s show that the iron railings, the iron bandstand and some herbaceous flower beds were removed, and in their place art nouveau reinforced cement benches were introduced (Silva 2010, 121–124) and several curved paths were created, modifying the layout of the square.

In 1937, the square's landscape was entrusted to Burle Marx who proposed building a public dance floor, an aquarium and a restaurant, although none of these came to fruition. His idea was to bring together the traditional garden's usefulness and a new recreational dimension, introducing "necessary attractions"

Figure 1.7 Praça da República (República Square): trees, art nouveau benches, imperial palm trees and curved pathways.

Source: Museu da Cidade do Recife, published in Silva (2010, 123).

such as dance floors where city dwellers could enjoy the music broadcast on the radio or played by a band (*Diario de Pernambuco* 1937).

The aquarium (featuring 20 species of fish) required the demolition of the building of the State Treasury, because it stretched until the waterfront, taking advantage of the presence of the river (*Diario de Pernambuco* 1937). In this way, the aquarium reflected the visual and conceptual integration of the garden with its surroundings, framed by the Capibaribe River. In addition, Burle Marx conceived of a fountain with lighting effects to occupy the centre of the square (where the bandstand had been), producing scenic effects and highlighting the monumental character of the garden and of the historic site.

Some other aspects of the garden were changed: the art nouveau benches were replaced by polished granite benches, the former layout of cruciform paths and curved paths was redefined, and diagonal paths were introduced to emphasize the monumentality of the architectural ensemble.

Burle Marx decided to preserve some of the existing nineteenth-century statues depicting figures from classical mythology "because they represent something interesting" (*Diario de Pernambuco* 1937). As for vegetation, despite the absence of a detailed survey of the existing plants at the República Square prior to Burle Marx's reform, an inventory based on historical, iconographic and documentary sources was prepared and allowed for the identification of some species

Figure 1.8 Praça da República (República Square): granite benches, imperial palm trees, fountain with lighting effects and diagonal pathways, *c*.1941. On the left is the Saint Isabel Theatre.

Source: Museu da Cidade do Recife, published in Silva (2010, 172).

that were introduced between the nineteenth century and the beginning of the 1930s (Silva 2016). Among them were the exotic areca palm (*Dypsis lutescens*), baobab (*Adansonia digitata*), fícus-benjamim (*Ficus benjamina*), flamboyant, phoenix palm (*Phoenix roebelenii*), imperial palm (*Roystonea oleracea*), Chinese palm tree, sabal palm (*Sabal palmetto*), pine (*Araucaria* sp.) and ravenala (Silva 2016, 383). Burle Marx kept some of these tree species, such as the *baobab*, the *fícus* and the palm trees, notably the imperial palm tree, which had been widely cultivated in Brazil. The landscaper explained that the old palm trees were out of line, but they would "only be removed when the new ones are in their 20s, able to replace them" (*Diario de Pernambuco* 1937).

In the words of Joaquim Cardozo (*Diario da Tarde* 1937), Burle Marx's project reconciled "the complete remodeling of the physiognomy of that place" with the "use, for decorative purposes, of the main natural elements that were present in its primitive form". Cardozo also stressed that some of the most beautiful Brazilian palms, such as *macaíbas* (*Acrocomia intumescens*) and coconut, as well as trees "so despised in Pernambuco" such as the cashew tree (*Anacardium occidentale*) and the *mangabeira* (*Hancornia speciosa*), were planted for the first time in the República Square (*Diario da Tarde* 1937) without "destroying the figs and the magnificent *baobab*" (*Diario da Tarde* 1937). The *macaíba* or *macaúba*, an autochthonous palm tree that "so well characterizes the landscape of the surroundings of Olinda and Recife" (Marx 1985, 71), had already been used in the Casa Forte Square. As a whole, the choice of plants aimed to bring together harmoniously old and new, artificial and vegetal elements: "without destroying the old, Mr. Burle Marx will add something more, adding a new and broader meaning to the ongoing reform" at the República Square (*Diario da Tarde* 1937).

In his analysis of Burle Marx's projects for both the squares of Casa Forte and Euclides da Cunha, Silva (2017, 151) highlighted the "relevant role played by native vegetation and the valorization of naturalized [species] which were already part of the landscape of Recife". This is also true of other Burle Marx projects, such as the República Square, where the landscaper kept the imperial palms, already acclimatized to Brazil and in line with the historic character of the garden.

A *Praça do Derby* (Derby Square)

Another important work by Burle Marx was the Derby Square renovation, begun in 1937 but which remained unfinished, because the landscaper left the post and the city of Recife for political reasons.

Derby Square, also called Derby Park, was built as part of a new residential neighbourhood project, which included a canal for drainage, public roads and houses. Derby neighbourhood was built in a vast area which had formerly hosted a hippodrome and, later on, a commercial and recreational complex belonging to the businessman Delmiro Gouveia, and which included a market, a hotel, a velodrome, a casino and an amusement park. Gouveia's property was destroyed in 1900 by a politically motivated arson attack, leaving behind a large open area called *Campina do Derby*.

In the 1920s, the Derby neighbourhood project was undertaken by Governor Sergio Loreto and Mayor Antônio de Góis, as part of a wider plan of urban works that included public gardens. Inaugurated in 1924 in *Campina do Derby*, the new square was divided into two parts by a wide avenue that connected the nearby neighbourhoods to the police headquarters that used the building of the old market. The project included two pergolas – one in the northern part and the other in the southern part of the square – in reinforced cement for climbing plants, a small lake, a lake with an artificial island called *Ilha dos Amores* (Island of Love) and two types of art nouveau benches in reinforced cement. Thus, each main area of the square contained a building and a water feature; the walkways among these areas were rectilinear, and incorporated benches and trees to encourage visitors to sit and enjoy the park.

As in the previous case, there is no detailed survey of the plants that existed at Derby Square prior to Burle Marx's work. However, the species introduced before the 1937 remodel were the *fícus-benjamim*, the flamboyant, the imperial palm, the red-lucky seed (*Adenanthera pavonina*), (*Livistona* sp.) and the tamarind (*Tamarindus indica*) (Silva 2016, 383–384).

Figure 1.9 Praça do Derby (Derby Square): art nouveau benches, ornamental lake and trees.

Source: Museu da Cidade do Recife, published in Silva (2010, 100).

Burle Marx's plan included an extensive list of vegetation but did not account for the species that already existed in the square. In his writings, the landscaper mentioned native and exotic plants as part of a global strategy to build a garden with "exclusively tropical material" both by using "an immense variety of plants offered by our magnificent woods" and by using exotic plants that "have already fully adapted to our climate" (Marx 1935b, 1). He mentioned that "a collection of 70 species of palm trees brought from the Rio [de Janeiro] Botanical Garden would be used" at Derby Square (*Diario de Pernambuco* 1937). An enlarged version of the Island of Love – Marx considered that its charm "loses in its miniature. It gives us the idea of 'artificial', since nature is vast and requires, for its imitation, larger dimensions" (*Diario de Pernambuco* 1937) –

Figure 1.10 Praça do Derby (Derby Square): granite benches and palm trees, probably redesigned by Burle Marx.

Source: Fundação Joaquim Nabuco, published in Silva (2010, 177).

would host purple orchids (*Orchidaceae*), purple castor beans (*Ricinus communis*), coastal morning glory flowers (*Ipomoea cairica*) and mangrove (*Diario de Pernambuco* 1937). Few documents and photographs are available to confirm whether all these species were actually planted on the Island of Love.

Burle Marx redesigned both lakes and introduced benches of polished granite; the pergolas, which are still in place, were not draw in the plan, a probable indication that he intended to remove them at a later stage. According to the plastic artist Francisco Brennand, who was interviewed by the local press in 2000, Burle Marx "said that he was working at Derby Square at the time he was accused of being a Communist. They accused him of digging the square to make trenches" (Alves 2000, 10).

Burle Marx's designs for the República Square and Derby Square took into account the elements already *in situ*, the plasticity of the plants and the interpretation of the character of the local landscape. At the República Square, the expressiveness of the surrounding water, the monumentality of contiguous buildings, the solemn and institutional aspect of the surroundings, the historicity of the garden and the place where it was located were the guiding elements of the intervention. At Derby Square, the garden's scale – larger than an ordinary square and compatible with an early twentieth-century urban park – together with the fact that it was a recent project encircled by a residential area pointed to the use of numerous and diversified species in combination with small, pleasant and relaxing spaces that invited residents to use and explore it.

Conclusion

Upon assuming a post in the Government of Pernambuco at the age of 25, young Roberto Burle Marx designed the first public gardens of his landscaping career, setting the pace for the new Brazilian modern gardens anchored in his artistic skills and botanical interests.

Burle Marx experimented with new ideas and concepts for shaping landscapes, such as an understanding of the local landscape and the gardens' surroundings, respect for the human element, and the use of vegetation as the main protagonist in both the creation and use of public gardens.

In his first projects, he designed themed gardens based on Brazilian native plants as well as tropical plants from different geographical regions. He used a large and diversified set of native species, valuing the biomes of the Amazon and the Caatinga, perceived as a harmonious ensemble of humans and non-humans (hence the role played by the statues which portrayed the native inhabitants of Brazil's inland). Later on, in other garden projects, he used pre-existing vegetation, native and exotic species already acclimatized to Brazil, common in city gardens, and also preserved existing historic artefacts and features.

Although respecting nature and campaigning for an ecological approach when designing his gardens, Burle Marx still perceived them as a sign of human agency and control over nature, as expressions of art. Gardens were architectural human-driven creations, as well as spaces that fulfilled specific functions related to hygiene, education and art; they were designed to be inhabited by human beings and every element was conceived to provide visitors with well-being conditions and recreational facilities: the shadows of trees, the architectural effects of water, the pioneering use of water as a means for cultivating aquatic plants, the walkways along the different ensembles of plants. In a nutshell, Burle Marx's gardens were thought to be part of the *civitas*, thus contributing to reinforce the bonds supporting the collective body of all citizens.

Acknowledgements

I wish to thank Joelmir Marques da Silva for help in identifying the plants and their scientific names.

Notes

1 Recife is the capital city of the state of Pernambuco located at 08° 03'14" south latitude. It has a hot and humid climate and the air temperature varies little throughout the year. The abundance of water explains the birth and development of the city strongly shaped by rivers and bridges. Often known as the "Brazilian Venice", Recife is characterized as an aquatic city or, using the words of the Brazilian physician and writer Josué de Castro, as an "amphibious city".
2 In 2015 six public gardens/squares – *Praça de Casa Forte* (1935), *Praça Euclides da Cunha* (1935), *Praça da República* (1937), *Praça do Derby* (1937), *Praça Ministro Salgado Filho* (1957), *Praça Faria Neves* (1958) – and the Garden of the Palácio do Governo de Pernambuco were classified as national heritage sites. In 2016 those seven gardens among others were protected by municipal law (*Jardins Históricos de Burle Marx*).
3 This text was reformulated and expanded and then republished under the title "O Jardim da Casa Forte" (The Casa Forte Garden) in the newspaper *Diario da Manhã* on 22 May 1935. Parts of the same text were also published in the report "A *reforma dos jardins publicos do Recife*" ("The reform of the public gardens of Recife") in the newspaper *Diario da Tarde* on the same day.
4 The Amazon is the largest tropical forest in the world, covering around 600 million hectares. More than half of the area of this biome is in Brazilian territory. It includes the Amazon rainforest and the Amazon basin (the largest in the world), fed by the Amazon River. The equatorial climate of the Amazon presents high average temperatures (22° to 26°C) and high average rainfall (2,300 millimetres).
5 "Indian" is a colloquial expression for naming the native inhabitants of Brazil before being colonized by Portugal. In more formal contexts the terms "native American" and "aboriginal" are used instead.
6 This taxonomic identification is based on the specimens observed in Burle Marx's drawings and in photographs of the 1930s and 1940s, the list of species in the original plan of Casa Forte Square, and newspapers and official reports.
7 The word "*sertão*" has several meanings. Here it refers to the dry, semi-arid area of the inland of the northeast of Brazil. Its natural and human characteristics were portrayed by Euclides da Cunha in his book *Rebellion in the Backlands* (Chicago, IL: University of Chicago Press, 1944, English translation).
8 Caatinga is a Brazilian biome characterized by a type of desert vegetation and a semi-arid, hot and dry climate. The name "Caatinga" is a Tupi-Guarani word meaning "white forest" or "white vegetation", referring to the colour of plants during drought conditions.

Bibliography

Adam, William Howard. 1991. *Roberto Burle Marx: The Unnatural Art of the Garden* (catalogue of exhibition). New York: The Museum of Modern Art of New York.

Alves, Cleide. 2000. "O Recife de Burle Marx". *Jornal do Commercio*, 17 September. Cidades.

Annuario de Pernambuco. 1935. *Resumo estatistico e descriptivo das actividades pernambucanas em seus varios aspectos, 1935*. Recife: Officinas do Diario da Manhã.

Brito, F. Saturnino Rodrigues. 1917. *Saneamento de Recife: Projecto de melhoramentos*. Estado de Pernambuco-Brazil.

Cardozo, Joaquim. 1973. "A Diretoria de Arquitetura e Urbanismo (DAU): olhada de um ponto de vista atual". In *Forma estática-Forma estética: Ensaios de Joaquim Cardozo sobre Arquitetura e Engenharia*, edited by Danilo Matoso Macedo and Fabiano José Arcadio Sobreira (pp. 171–176). Brasília: Câmara dos Deputados: Edições Câmara.

Costa, Francisco Augusto Pereira da. 2001. *Arredores do Recife* (2nd edition). Recife: Massangana.

Diario da Tarde. 1935. "A reforma dos jardins publicos do Recife". 22 May.

Diario da Tarde. 1937. "Jardins bonitos que o Recife possue". 14 June.

Diario de Pernambuco. 1937. "*A reforma dos jardins do Recife*". 20 May.

Eliovson, Sima. 2003. *The Gardens of Roberto Burle Marx*. Portland, OR: Timber Press (Foreword by Roberto Burle Marx).

Ferreira, Domingos. 1932. *Planta da Cidade do Recife e Arredores*. Recife: Prefeitura Municipal.

Hamerman, Conrad. 1995. "Burle Marx: The Last Interview". *The Journal of the Decorative and Propaganda Arts*, 21: 156–179. https://doi.org/10.2307/1504137.

Jornal Pequeno. 1935a. "Ontem, Hoje e Amanhã". 9 July.

Jornal Pequeno. 1935b. "Ontem, Hoje e Amanhã". 31 August.

Jornal Pequeno. 1935c. "Ontem, Hoje e Amanhã". 16 October.

Jornal Pequeno. 1935d. "Ontem, Hoje e Amanhã". 31 October.

Jornal Pequeno. 1936. "Ontem, Hoje e Amanhã". 3 March.

Marx, Roberto Burle. 1935a. "Jardins para o Recife". *Boletim de Engenharia*, 7(March).

Marx, Roberto Burle. 1935b. "Jardins e Parques do Recife". *Diario da Tarde*, 14 March.

Marx, Roberto Burle. 1935c. "O Jardim da Casa Forte". *Diario da Manhã*, 22 May.

Marx, Roberto Burle. 1954. "Conceitos de Composição em Paisagismo". In *Arte e Paisagem: Conferências Escolhidas*, edited by Roberto Burle Marx (pp. 11–19). São Paulo: Nobel.

Marx, Roberto Burle. 1962. "Projetos de paisagismo de grandes áreas". In *Arte e Paisagem: Conferências Escolhidas*, edited by Roberto Burle Marx (pp. 51–60). São Paulo: Nobel.

Marx, Roberto Burle. 1985. "Minha experiência em Pernambuco". In *Anais do Seminário de tropicologia: Homem, terra e trópico*, edited by Maria do Carmo Tavares Miranda (pp. 68–86). Recife: Fundaj/Editora Massangana.

Melo, Mario. 1935. "O caso do Jardim da Casa Forte". *Jornal do Commercio*, 26 May.

Montero, Marta Iris. 2001. *Roberto Burle Marx: The Lyrical Landscape*. Berkley; Los Angeles: University of California Press.

Prefeitura do Recife. 2016. *Decreto no. 29.537*. Recife: Prefeitura Municipal.

Sá Carneiro, Ana Rita. 2017. "Quinta porta: o projeto do jardim como paisagem". In *Cadernos de Arquitetura e Urbanismo: Cidade-paisagem*, edited by Lúcia Veras *et al.* (pp. 78–95). Recife: CAU/PE, João Pessoa: Patmos Editora.

Sá Carneiro, Ana Rita, Aline de Figueirôa Silva, and Joelmir Marques da Silva (eds). 2013. *Jardins de Burle Marx no Nordeste do Brasil*. Recife: Editora Universitária da UFPE.

Silva, Aline de Figueirôa. 2010. *Jardins do Recife: uma história do paisagismo no Brasil (1872–1937)*. Recife: Cepe.

Silva, Aline de Figueirôa. 2016. "Entre a implantação e a aclimatação: o cultivo de jardins públicos no Brasil nos séculos XIX e XX". PhD dissertation, Universidade de São Paulo.

Silva, Aline de Figueirôa. 2018. "Brazilian Gardens in Historical Perspective: Notes on the Origins, Functions, and Design of Recife's Squares". *Patrimônio e Memória*, 14(1) (Unesp): 111–125. http://pem.assis.unesp.br/index.php/pem/article/view/828.

Silva, Joelmir Marques da. 2017. "Integridade Visual nos Monumentos Vivos: os jardins históricos de Roberto Burle Marx". PhD dissertation, Universidade Federal de Pernambuco.

2 Between the nuclear lab and the backyard

Artificially enhanced plant breeding and the British Atomic Gardening Movement

Vanessa Cirkel-Bartelt

Introduction

When Muriel Howorth, British society lady, garden and science enthusiast and biographer of Frederick Soddy, founded the British Atomic Gardening Society in 1960 with the help of scientists and professional gardeners, she had a clear aim in mind. Hobby gardeners should participate in radioactive breeding by sowing the irradiated seeds into their backyards, by monitoring the results and by reporting any significant changes to scientists. Thus, the development of new plant mutations should be improved.

What looks like an anecdote from garden history raises numerous interesting questions. What role did radiation play in the artificial enhancement of plant breeding, compared, for example, to chemical methods? Was it a rarely used method or relatively common? There are hints that a significant number of mutations bred in this way found their way into our supermarkets.[1] So what about the Atomic Gardening Movement? Was it happily received as a means to breed more profitable, healthier or more nourishing types of plants? Howorth stated that she saw radioactive breeding as a chance to stop worldwide famine (Howorth 1960, 4–5), but it is questionable whether her enthusiasm was mirrored by the actual practitioners. Did the gardeners perceive the irradiated seeds as something unnatural? We are all well aware of the ongoing debates about genetically modified (GM) plants, but seeds from irradiated plants seem not to have been met with such strong, let alone violent, protest as, for example, the "gene maize". How did scientists and laymen react to the artificially treated seeds, especially in the context of the garden?

The context of gardening also poses an interesting framework to ask about what actually defines what we call "the Anthropocene". Gardens have always been under the immediate influence of man – spheres where nature and culture become an amalgam and thus form something entirely new. So what distinguishes "normal" human interference with nature from its anthropogenic manifestation? Analysing the historical development of Atomic Gardening and its scientific parent technology – radioactive[2] breeding – opens up a new perspective on the relationship between nature and culture, as it took the manipulation of plants to a whole new level.

Traditionally, critique of technology, be it in its specific philosophical shape or in historiographic reflection, used to rely on the dichotomy between "nature" and "culture", focusing in particular on the cultural aspects of science and technology. More or less based on Marx's theory of alienation, this form of critique presumed that a kind of ideal primordial natural state existed, but humankind could no longer reach it, since it depended too heavily on technology (Klems 1988, 118–166). This conventional division was challenged in the 1960s by new approaches, such as Boulding's "Spaceship Earth". He applied the concept of the closed system to the ecosphere and thus influenced to a large degree the debates about ecology in the second half of the twentieth century (see Boulding 1966). More recently, Crutzen and Stoermer's idea of the Anthropocene has also led to a more systematic view of the interaction between human beings and technology on the one hand and nature or the environment on the other (see Crutzen and Stoermer 2000; Trischler 2018). These approaches are quite useful for analysing systematic correlations, for example, between man-made CO_2 emissions and climate change. In addition, focusing on the effects of human action on the environment, they make it desirable to explain the historical evolution of certain groups of objects that are natural and at the same time have been influenced, altered or modified by human intervention. These objects, like cultivated plants, in a way are "natural", since they are living, not man-made in the original sense, and are not jointed from various materials. And yet, just like any technical artefact, they would not exist – at least not in their current shape – without human action. Karafyllis has coined the term "biofact" for these natural-technical hybrid objects (see Karafyllis 2006).

Yet, the concept of the biofact alone is not enough to explain all types of human interactions with natural objects like plants. Russell suggests writing the history of the development of objects like plants and animals as an "evolutionary" history that not only considers the natural factors that may influence the evolvement of certain species, but also the aspects from cultural history and their mutual interdependency with nature (Russell 2004, 1–2). The progress of induced mutagenesis by irradiation as a means of plant breeding has been influenced by various cultural trends that surpass its entanglement with radioactivity research. It is very likely that the reason why the method of radioactive breeding was and is so readily accepted lies in the fact that it promised to combine the benefits of technological scientific plant breeding with the more general hopes of a better living through technology. Atomic Gardening enthusiasts and scientists alike aimed at mechanizing the breeding process, anticipating the promise of costume-designing organisms that would finally be fulfilled with the advent of genetic modification. The garden enthusiast's approach also transcended the usual chain of production, where the plant is bred by experts and consumed by the layperson. In Howorth's movement – if it had ever come to full fruition – every hobby gardener would also have been a hobby scientist with the status of a voluntary lab hand.

So how may a historical understanding of radioactive breeding add to our understanding of the Anthropocene? One of the main difficulties of the concept is the question of periodization – where does one draw the line? For, in a way, developments such as the Neolithic Revolution may also be understood as the beginning of a new geological age defined by human beings tampering with nature (McNeill and Engelke 2014, 1). Yet, in a strict sense Anthropocene refers to more recent times, usually from the Industrial Revolution onwards. This caesura is motivated by the fact that anthropogenic effects, such as global warming, seem to have accelerated probably since the mid-nineteenth century and seem even to have increased following World War II (McNeill and Engelke 2014, 4–5).

Radioactive breeding seems to be a good example of these accelerated alterations of nature. Russell introduced the idea of human beings being able to transform natural objects wilfully as much as accidentally and that the development from "wild" organism to biofact is actually a continuum (Russell 2010, 255). He classified these transformations into five groups of methods that human beings used historically to interfere with nature: capture, taming, domestication, breeding and genetic engineering (Russell 2010, 255). According to this model, radioactive breeding bridges the gap between "normal" breeding and genetic engineering.

In what follows, this chapter will outline the British Atomic Gardening Movement and ask if and how far a similar movement may have existed in other countries. In addition, the scientific background of radioactive breeding will be explained. The chapter will then turn to the question of how these methods were received outside the scientific community and particularly why they did not meet with much protest. The chapter will address the question of whether and how these breeding procedures are linked to the Anthropocene.

How does radioactive breeding work?

As the name already indicates, radioactive breeding is achieved by the irradiation of plants, in particular their seeds or the fruits that produce the seed. The source of radiation is usually a gamma-ray, like the artificially produced cobalt radioisotope Co60. The seeds of those plants that have not been burned by the radiation[3] will then be made to sprout – that is, of course, if the procedure has not rendered the plant sterile. Afterwards, the seedlings will be picked and planted as would "normal" plants. In most cases the plants of the second generation do not differ from their mother plant. In rare cases, though, there may be changes in colour, size or form. Before the advent of modern genetics the second-generation seedlings had to be closely monitored in order to find such divergences. Those plants showing altered characteristics can thus be used as a basis to breed a new variety (Shu and Forster 2012, 9). This kind of "trial-and-error" mutation is called "induced mutagenesis" (Shu and Forster 2012, 11) and differs from methods like DNA sequencing by the randomness of the occurring changes. Induced mutagenesis can also be achieved by chemical treatment, the best known being an alkaloid named colchicine (see Curry 2016).

The British Atomic Gardening Movement

The activists of the Atomic Gardening Movement planned to provide gardeners with mutagenic seeds. "Atomic blasted seeds" had been bred and were branded in the USA. The Atomic Gardening Society finally imported them to the UK (Howorth 1960, 5).

One reason for the interest in plants bred by radioactive means may have been the general interest in new plant varieties in industrialized countries. Since the mid-nineteenth century gardening and gardening supplies have become an important economic factor. Ornamental plants became a modern consumer product and, as Lanman has shown, the right breed of plant could turn a nursery into a successful enterprise (Lanman 2004, 34–36; De Vries 1905, 813). Flowers for hobby gardeners and professional florists had to be adapted to the changing fashions. Variations in colour, size or other properties are now dependent on the varying tastes of consumers (Lanman 2004, 37). Thus, to a certain degree, radioactive plants where there to answer the need of the market, and the estimated value of a new radioactive variety could reach up to US$50,000 (Curry 2016, 183). But, apart from that, "atomic" plants were not only supposed to enrich the repertoire of plants for hobby gardeners, but those gardeners should also be turned into hobby scientists, a trend that had been established in the nineteenth century when science started to gain importance in non-professional gardening (Curry 2014, 547). Modern-day professional horticulture itself seems to have developed in many countries through close cooperation between commercial and scientific actors (see e.g. Duarte Rodrigues and Simões 2017).

Induced mutagenesis, the key method behind atomic plants, is prone to random alterations in contrast to site-directed mutagenesis. It is never safe to say which genes will change due to the procedure and what effect this may have on the plant's properties. Today, GM screening methods help to find altered gene sequences. Back in those days, seeds that had been produced by irradiated plants had to be made to sprout and planted, and had to be monitored in order to find out if the resulting seedlings showed any signs of change. The growing and monitoring process is quite time consuming, so the hobby gardeners should step into the breach. They should enlarge the acreage for irradiated plants by making their own gardens available as a kind of test garden. They should also take over the monitoring tasks. The atomic gardening enthusiasts' credo was: "More mutation experimenters, more quickly to find more food for more people" (Howorth 1960, 69). The mention of the need to find higher yielding plants for food production was owed to older *topoi* about how to fight world hunger. But the atomic gardeners under the aegis of Muriel Howorth would also advertise their cause by hinting at the opportunity of gaining fame by naming newly discovered varieties (Howorth 1960, 51).

The British Atomic Gardening Society was founded by Muriel Howorth in 1960. During World War II she had been working for different ministries, her main task being the production of information materials (Johnson 2012, 552). After the war she had been asked to focus on atoms, and over the following years she published a series of booklets that described the peaceful use of radio-active substances and atomic energy in a simplified and fun way for different lay audiences (see e.g. Howorth 1955). But Howorth also took a private initiative. She founded a number of clubs and societies, and hosted various events in order to inform an educated audience about the different uses of the atom (Forgan 2003, 188).

One of these initiatives was the Atomic Gardening Society, founded in 1960. The society was supposed to introduce non-scientists to the background of radio-active breeding and help them perform their tasks as volunteers for the afore-mentioned plant monitoring. According to Howorth, she had a waiting list of 300 people who wished to become a member of the society (Howorth 1960, 16). The society (i.e. Howorth) published a guidebook that gave a detailed account of how to proceed if you wanted to become an atomic gardener. First you had to establish a "laboratory" garden for the radioactive plants and a control garden (Howorth 1960, 55). There was also a step-by-step account of how to work with the irradiated seeds (Howorth 1960, 53–57):

1 Planning (what do you want to see in your garden?)
2 Indoor planting
3 Transplanting
4 Outdoor planting (with an experimental and a control garden)
5 Finding mutations
6 Pollinating changed plants
7 Stabilizing the mutation.

The book would also offer general tips for the gardeners (e.g. how to prepare the soil). The Atomic Gardening Society offered forms for data sheets and gave instructions so that its members would be able to report their mutations prop-erly (i.e. in scientific way) to the society (Howorth 1960, 55–56).[4]

Apart from administrative responsibilities, the society also organized visits to parks and public gardens that had atomic plants on display (e.g. the Royal Botanical Gardens in Kew) (Howorth 1960, 65). So far there has been no historical investigation into the question of how common such exhibitions of plants bred by radiation-induced mutagenesis actually were and whether they had been on display in public spaces in other countries in Europe as well. The known British examples at least hint at a certain public interest in the results of this method. Furthermore, Howorth and other members of her society (e.g. the radiochemist Frederick Soddy) also exhibited the atomic plants they had culti-vated in their own gardens at prestigious events such as the Royal Horticultural Society's annual garden show. There the atomic plants stirred considerable interest and their display was covered by the media (Howorth 1960, 15–18). Some reactions to the garden enthusiasts' endeavours were rather negative. Audiences of the garden show, for example, asked why the matching specimen from the control garden was not also on display so that visitors could judge the differences between normal and atomic plants with their own eyes. There is no evidence, though, that the criticism was sparked by a techno-critical point of view in general. Only one person is reported to have remarked that the atomic gardeners where touching on a hot topic and accused them of "using public concern with nuclear dangers as a means of securing free advertisement" (Howorth 1960, 18).

Atomic gardens in America and abroad

As Muriel Howorth was the head of the movement, her old age and deteriorating health made the British Atomic Gardening Movement a short-lived enterprise. Only a few years after its foundation, she had to give up her activities. Unfortunately, not much of her private correspondence in connection with atomic gardening seems to have survived (Johnson 2012, 564). Existing sources such as the aforementioned booklet suggest that before 1960 there was not much of an atomic gardening "movement" in other European countries. Howorth and her colleagues considered themselves to be pioneers when it came to introducing radioactive treated seeds. Nevertheless, according to Howorth, there were individual gardeners abroad who had joined her movement (Howorth 1960, 24). The ever industries science popularizer, Howorth seems to have planned to tackle this issue and garner more international cooperation with the planned future foundation of "The World's Atomic Experimental Gardens" (WAEG). A list of potential member countries very generously mixed countries and whole regions, including "England, France, Greece, Sudan, Rhodesia, Malaya, Formosa, Hong Kong, Solomons, Australia, South America, Haiti, Canada, America, Ireland, Scotland".[5]

More interestingly, those European laboratories which had already succeeded in breeding new varieties by radioactive irradiation seem to have been reluctant to hand over their material to laypeople and even the British plant breeders seem to have been reluctant to accept Howorth's help (Howorth 1960, 47). This was probably the reason why Howorth turned to a certain Dr Speas in order to import irradiated seeds from the USA. Even more astonishingly, Sweden and other countries where radioactive-bred plants already existed (Howorth 1960, 69) are not included on the list for potential WAEG members – a hint that the scientific breeding community was obviously not too eager to share their knowledge with an interested public.

The irradiated seeds were sold by a small private US company called Oak Ridge Atom Industries Inc. led by Dr Speas,[6] who seems to have cooperated with the National Oak Ridge Laboratories. The laboratories' scientists provided him with a concrete casing for his radiation source and a source of "ten curies of cobalt-60" (Howorth 1960, 19). Speas had been granted an official licence in 1957 allowing him to irradiate plants and seeds (Howorth 1960, 19). Atom Industries Inc. not only sold seeds to aspiring gardeners but also offered a starter kit for educational purposes, hoping to get school classes involved in the search for mutations (Howorth 1960, 26). Speas had found a fellow campaigner in Muriel Howorth and he seemed to have made similar efforts to inform lay audiences in the USA about atomic gardening, by promoting it at garden shows and telling his story to the media. The commercial success seems to have been significant enough that other companies tried to trade irradiated seeds as well (Curry 2016, 182–185).

Radioactive breeding and its beginnings

While radioactive breeding prospered from the mid-twentieth century onwards, its history can be traced back to the beginnings of radioactivity research. The turn from the nineteenth to the twentieth century was not only the heyday of various new types of physics, but similar developments also took place in biology and its growing number of subdisciplines. Based on the rediscovery of Mendel's ideas about heredity, other early geneticists like Hugo De Vries established the concept of spontaneous mutation as the biological explanation for the mechanisms that enable the evolution of species and their adaption to ecological niches. According to Curry, De Vries was also the first to suggest the use of ionizing radiation to induce mutations artificially (Curry 2016, 17).

Over the following decades scientists in Europe and the USA started to experiment with the irradiation of plants and lower organisms. Systematic analyses of the early biological experiments are still pending, but there are cases of single institutions and people involved in radioactivity research. In 1912 scientists from the Radioactivity Laboratory (Laboratorio de Radiactividad) of the University of Madrid conducted the earliest research regarding the effects of radiation on plants.[7] They enriched substrate with thorium as a kind of fertilizer and observed its influence on the growth of the plants in comparison to specimens that had not been treated with radioactive substances (Herrán 2008, 142f.). The Spanish researchers noted that the treated plants grew more regularly and became larger (Herrán 2008, 143, image 6.3). As a consequence, the head of the institute, José Muñoz del Castillo, tried to establish the *agricultura radioactiva* for a broader audience. He gave public lectures aimed at educated laypeople such as agricultural engineers (Herrán 2008, 145).

One of the most important European institutes in radioactivity research was the Radium Institute (Radiuminstitut) in Vienna.[8] Founded in 1910, the first tests of how radiation may influence organic matter were started the following year. Edgar Congdon, a visiting American geneticist, worked on the effects of radiation on larvae and fly eggs (AÖAW 1911), while the head of the local Institute of Biology, Hans Molisch, analysed the effects on plants (AÖAW 1911; Knierzinger 2012, 112). Molisch continued his work on radioactivity and its effects at least into the 1930s when he worked for the Research Institute (Forschungsinstitut) Gastein. There he analysed the development of microorganisms and plants in thermal water containing radon (Knierzinger 2012, 117). Thus far, the biological works conducted in research institutions dedicated to radioactivity research have not been historically analysed.

Works on the history of the Cavendish Laboratory at the University of Cambridge – the United Kingdom's leading institution in early radioactivity research – suggest that its researchers did not work on plant irradiation before the 1930s (see e.g. Kim 2002). Nevertheless, the idea that radioactive substances may be of future use for food production, by using it as a kind of artificial fertilizer, was already well established in public opinion in the 1920s. When the Vienna Radium Institute lent a large amount of radium to the British government in 1921 newspapers were full of articles about fantastic stories of how radioactivity might be used in science, medicine and agriculture (Soddy 1921).

The late 1920s saw other important achievements in the development of the induced mutation method. Herrmann Joseph Muller managed to alter the genetic set-up of drosophila flies by exposing them to X-rays (Muller 1927, 84–87). Muller and other geneticists did not make use of this approach in order to alter organisms in hindsight of future economic use. What they were actually looking for was the secret of the genes. The experiments with modified flies were supposed to prove or falsify theories about the properties of genes in general (Muller 1927, 86; see also Kohler 1994). Besides, the solving of this puzzle promised to answer open questions about how organic matter evolved (Falk and Lazcano 2012, 374).

Swedish geneticists tried to prove that Muller's approach would also work in the case of plants and started to treat barley with X-rays as well as ultraviolet radiation in the late 1920s (Lundqvist 2014, 123). They seem to have been the first to experiment with the use of Co60 as a source of high energetic gamma-rays (Howorth 1960, 11). It took about a decade, but finally the induced mutagenesis led to the production of a new variety of Swedish barley (Lundqvist 2014, 123).

Other nuclear agricultural technologies

Probably because the knowledge about the chromosome set was not sufficiently developed for site-directed mutagenesis (i.e. the purposeful alteration of organisms), scientists soon found additional areas of application for radioactive substances in the production of agricultural crops and food.

One of these applications was the use of so-called "tracers". Radioactive fluids are injected into plants in order to help discover how fertilizers or nutrients are dispersed throughout the plants' organisms. The more important use of radioactivity in crop and food production was the radiation of plants, certain parts of plants and their fruit, as well as other food (e.g. meat and meat products). The method aimed to kill micro-organisms that spoil food, such as salmonella. Radiation was also used in order to render male insects sterile, so that vermin would not destroy crops and flies would not infest meat (Hamblin 2009, 30; Zachmann 2013, 10; 2012, 181–182). The irradiation of food was and still is highly controversial. In the USA in the 1950s it was mainly the food industry and particularly the producers of conventional frozen foods which tried to stop the application of that method, as they feared competition of cheap foods with a long shelf life that would last without the help of a freezer (Zachmann 2012, 183). In other industrialized countries (e.g. Western Germany), the radiation method was not a success at first, since it was not as reliable as freezing when it came to treating products like fish for preservation. Or rather, a successful treatment of critical foods would have been far too expensive, as the radiation technique consumes a lot of energy (Zachmann 2012, 191–192).

Without going into more detail about the use of radiation and how people reacted to the method in different countries, it is worth mentioning that food irradiation was often used in a postcolonial context following World War II. The effective application of nuclear agricultural technologies was intended to leave a good impression of the "peaceful atom" and thus pave the way for other peaceful uses (e.g. in the case of nuclear energy) (Breitwieser and Zachmann 2017, 113–115).

Protest and lack thereof

Modern genetic engineering and the production of GM plants especially for food and forage have caused considerable public outrage, particularly in industrialized countries. Critics of GM plants have two main arguments. First, GM plants may cross-breed into "wild" varieties of plants. Being better adapted to the very environment, they may thus threaten to replace the native varieties or even whole species. There is also the imminent danger of the so-called "interspecific gene flow": GM characteristics of the new plants may be transmitted to other species. As a consequence, modified characteristics, such as resistance to certain chemicals, like herbicides or pesticides, could thus be transferred to the very pest or vermin they are supposed to fight, rendering these chemicals completely useless (Boyd and Prudham 2004, 128–129). Radioactive breeding on the other hand seems to have been completely ignored by contemporary protesters as much as by historians.

This is remarkable, since the fear of genetic changes as an unwanted side effect of radioactive radiation started to spread throughout the 1950s. As Weart has shown, it was in particular the fallout from the dropping of atomic bombs and nuclear bomb tests that raised public awareness of nuclear technologies, especially in the USA. Experts from the National Academy of Sciences of the United States were among the first to declare that any kind of radiation may have mutagenic effects in general (Weart 1988, 201). With this statement they opposed the official stance of the US government under the leadership of President Eisenhower. Moreover, this criticism evoked a new version of an old image: the scientist as Doctor Frankenstein, unable to control his own creation (Weart 1988, 48–49, 64–65). But in the case of radioactive breeding, trends developed differently, and in the late 1950s and early 1960s the method managed to achieve first successes and produced marketable new plant varieties (Howorth 1960, 69).

The method of radioactive breeding obviously never caused much protest, especially in comparison to other nuclear technologies that met so much resistance. It is important though to note that there never existed the one uniform reaction to the various types of nuclear technology at different times and in different places. The military use of atomic energy was met with public refusal in most Western countries. Still, it did not hinder the arms race between the East and the West during the Cold War. The public reception of the civil use of nuclear energy was even more ambivalent (Balogh 1991, 161–170). As Rucht has pointed out, even the situation in Central Europe was complex and neither the number of actual nuclear reactors in a country nor the density of the population provides a solid clue as to why in certain countries the resistance was much fiercer than in others (Rucht 1995, 277–292). The same seems to be the case with food irradiation: protests against the method changed considerably over the course of time. In the USA, for example, the negative attitude towards irradiation that prevailed in the first post-war decades did not prevent the reintroduction of its use in the 1990s. In particular, the irradiation of meat was not only relaunched but it was also widely advertised by the supermarket chains that sold the meat (Spiller 2004, 740).

The curious case of the British Atomic Gardening Movement and the way in which the method of radioactive breeding has been accepted is surely due largely to the optimistic view which many people in Western countries shared about the peaceful use of atoms. The speech "Atoms for Peace" given by US President Dwight D. Eisenhower on 8 December 1953 at the general assembly of the United Nations is usually considered to be the turning point for the peaceful use of nuclear energy. Eisenhower proposed using this resource "for the benefit of all mankind".[9] Nevertheless, not only was the speech a very cunning way to mention the US extensive nuclear arsenal threatening the enemy, but calling it the starting point for the Atoms for Peace Movement neglects the fact that the governments of many other countries had long since started to campaign for nuclear energy to no longer be associated with the cruelties of warfare (Goldschmidt 1997, 1–15).

The UK was one of the leading nations when it came to advertising the "peaceful atom". Since the 1940s an educated lay audience had been made familiar with the history of radioactivity research and the prospective use of nuclear energy, mainly through publications of books and booklets like the ones written by Howorth. In addition, a number of exhibitions introduced visitors to the topic in an entertaining way. The Festival of Britain in 1951, for example, featured atoms at three different exhibition sites, the central one in London using the imagery of Lewis Carroll's *Alice in Wonderland* – the "Atoms in Wonderland" display (Forgan 2003, 180–181, 188).

It is not quite clear how far governmental institutions had an interest in these private initiatives concerning irradiated plants, but it is not unlikely, since they promised the opportunity to study the long-term effects of radiation on plants. Given the rising number of nuclear tests in the late 1950s and early 1960s, such analyses may have been of indirect military use. Some of the official nuclear tests like the famous one of 1946 at the Bikini atoll were visited by geneticists who did research on the effects of the fallout on maize (Anderson *et al.* 1949, 639). If such a connection between plant genetics, scientific breeding and military radiation research exists, it may provide a fruitful basis for future historical analysis of the so-called "military-industrial complex" during the Cold War and its connection to environmental questions (McNeill and Unger 2010, 9–10).

Technology to the rescue

To understand the mainly positive reactions towards atomic gardening and radioactive breeding, it is worth taking a closer look at the hopes associated with this specific technology of plant breeding.

In his book about the fears and negative images associated with radioactivity and nuclear fission, Weart has shown that especially at the beginning of the twentieth century both had at first also been linked to positive visions for the future. Particularly important was the idea that "clean" energy from atoms may one day replace the fires and furnaces that were polluting the crowded cities in particular. The hope was that without the steam and the soot from conventional combustion, the future would only see "white cities" (Weart 1988, 8). As a result, the use of atomic energy was heralded as an alternative source of energy long before the discovery of nuclear fission in popular magazines and science fiction novels in many countries, particularly in Western Europe (Cirkel-Bartelt 2017, 40–47). But early radioactivity research not only gave new impetus to questions of energy. Its medical use has also been very important ever since radioactivity was discovered. Röntgen's works with X-rays may be the most famous application of gamma-rays, but the possible use of radiation as a cancer treatment was also discussed. The first attempts at experimental cancer therapy were even conducted before there was systematic research about the effects of radiation on bodies and cancer cells. It was Sievert who approached questions of dosage in his works in the 1920s (see e.g. Sievert 1923). As already mentioned, around this time the idea of using radioactive substances as a kind of fertilizer had also already been established.

Of course, it is well known that the use of products enriched with radium was absolutely in fashion in industrialized countries throughout the 1920s. From so-called radium "emanators" to produce radioactive water as a homemade cure, to cosmetic products, to medical items such as suppositories – innumerable lifestyle and health products had either been actually treated with radioactive substances or were at least advertising this claim (see Helmstädter 2006, 904–907). But this "radium hype" was surely not the only reason why an application of radioactivity to plant breeding was obviously received as a promising idea.

Fighting hunger

The production of plants that met the needs of industrialized agriculture was first and foremost due to the Industrial Revolution and the changes in agriculture (e.g. the introduction of steam-driven harvesters and similar machinery) throughout the nineteenth century.

But apart from that the development of professionalized plant breeding was also influenced by progress in other sciences as well. Now is not the time to go into detail about all the changes in agriculture, and in particular the economic revolutions that transformed the markets for agrarian products worldwide, especially throughout the late nineteenth century (Wieland 2004, 21–24), but one important consequence of these dynamics was not only the professionalization but also the scientification of plant breeding. Single initiatives by industrial food producers and farmers to improve plants were taken up by scientists and policymakers, and finally led to the formation of modern plant-breeding research (Wieland 2004, 18, 27–31, 147–150). Moreover, progress in biology and the genesis of its various subdisciplines also fostered the development of plant-breeding research. The evolvement of modern genetics, based on the application of Darwinian theories on living organisms and the "rediscovery" of Mendelian ideas about heredity, helped in particular to shape new concepts of plant breeding (Perkins 1997, 43–44, 54–58). New findings in chemistry, like the invention of artificial fertilizers, had an impact on agriculture as well.

These inventions and innovations were not only due to the scientific progress in natural sciences. Famine and food shortages were still a common problem in the late nineteenth century, even in Central Europe, but more so in colonial countries. The new scientific approaches were thus also aimed at gaining higher yields in crops in order to secure a stable food supply (Wieland 2004, 20). The progress in science and technology during the nineteenth century also altered the way famine was perceived. New findings, especially in thermodynamics, enabled scientists to calculate the exact energy needs of a city or even a country – be it the energy resources needed to keep engines running or the supply of an entire population with staple foods. The quantifiable facts fuelled the political debates that had been caused by the Industrial Revolution and the social problems that followed in its wake. No wonder that shortly after radioactivity had been discovered it was hailed as a remedy to all these ailments (Soddy 1903, 16–17).

The discourses about different aspects of energy all had a decidedly national perspective in common. William Stanley Jevons, who discussed the "Coal Question" (i.e. the future shortage of coal in Great Britain due to exhaustion of the resources), had already described dependency on foreign markets as a possible consequence of coal shortages (and deficit spending) (Jevons 1865, 17–21). William Crookes took a similar stance when he warned against the dwindling of national yields in wheat. In an address on the "Wheat Problem" he even considered food shortage and its economic and political consequences as a possible means of future warfare (Crookes 1898, 563). The dependency on corn imports was a problem in many countries worldwide (Perkins 1997, 48). Governments were no longer only concerned about importing plants and products of exotic origin that could not be grown in Europe (e.g. sugar or tea), but the focus shifted towards standard crops in order to provide an autonomous supply with resources, in this case, for food production. These attempts at autarky were to be topped by Germany during the time of National Socialism. Not only was much effort invested into the search for substitute materials (Heim 2003, 63–91), but also the aggressive policy of expansion was justified – among others – by the search for new agrarian resources (Heim 2003, 120–124).

Famine was still a problem in many countries even after the immediate hardships of World War II had ended. So the question of grain imports and the dependency of individual nations on these imports remained important. The regulation of grain prices also guaranteed a means of political regulation, particularly in colonial and postcolonial contexts. After the war this scenario was broadened by the dimension of the East–West conflict. Providing help with the production and provision of cheap staple foods was used (e.g. by the USA) as a possible means to improve living conditions in countries like Mexico, so that political unrest became less likely and the possible influence of communist parties was diminished (Perkins 1997, 119–120, 130–139; McNeill and Unger 2010, 6).

After the 1940s the various approaches to plant breeding led to a dramatic increase in different types of modified plants that were optimized for the soils they were grown on, the various harvesting methods or the yield they produced (Hamblin 2009, 37–40). Radioactive breeding became one of the most successful methods applied to produce these plants.

Conclusion

Radioactive breeding was and is a successful method to manipulate plants. Historically, it developed on the basis of different trends in nineteenth-century science (e.g. the progress in biology that led to the beginning of modern genetics), but progress was also made in chemistry and physics and finally economics that shifted focus on problems like worldwide famine.

Paradoxically, the success of science and technology in the nineteenth century to some degree even increased the problem of food crises, as it had accelerated (McNeill and Engelke 2014, 40–45) population growth and had indirectly – through the Industrial Revolution and its social effects – accelerated the pauperization of the working classes. In that respect, radioactive breeding, and in its wake also atomic gardening, are very typical offspring of the Anthropocene. Both were hailed as technological solutions to problems that would not have been as severe – it would be a little far-fetched to assume that famine would not exist at all – without the growing impact of technology on many spheres of life. In addition, and as a consequence, these agricultural technologies caused further problems that asked and still ask for more scientific or technological solutions (e.g. the unwanted side effects of cross-breeding with wild plants). This is probably also the reason why in comparison genetic engineering is received so negatively today: when this technology reached the markets, consumers had already developed a critical stance towards that downward spiral of man-made problems and "solutions" which often turned out to be new problems.

The particular interest in peaceful uses of nuclear technology during the Cold War era helped foster the method further. In particular, the attempts at introducing it to the sphere of the garden – public and private – seem to have established a lasting acceptance of radioactive breeding as a means to produce better plants. Moreover, the case of atomic gardening shows how science and technology depend on their popularization in order to have a lasting impact on society. The private initiatives of the early post-war years played a particularly successful role in this process.

Looking at the history of atomic gardening adds to our understanding of the Anthropocene as it directs attention to those "signals", for example, the CO_2 concentration in the atmosphere that provides scientific evidence for the existence of the Anthropocene as an actual geological epoch, but which is less often considered in public discussion in comparison.[10] In particular, the remains of a large number of so far non-existent plant varieties will leave traces in the strata of our planet. Thus, the case of radioactive breeding – similar to genetic modification – raises the question of long-term consequences in the unforeseeable future. Radioactive-bred plants may one day cause other plant varieties – or in the worst case scenario even whole species – to become extinct. This is a characteristic of our current understanding of the Anthropocene: the detected changes in the environment are considered unwanted side effects of human interference with nature. The case of atomic gardening reminds us that even in recent history this has not always been the case – science enthusiasts like Muriel Howorth happily embraced the changes they wrought in natural objects like plants as a potential benefit for mankind. So, any controversy about the Anthropocene and adequate responses to the problems associated with it will have to deal with defining what nature we actually want in the future and what types of interventions we collectively consider to be useful. Analysing human–environment relations on a small scale may also foster our understanding of how human interventions that usually take place at the local level and their side effects eventually added up – and still add up, causing the global changes in the Earth's system which we witness today.

Notes

1 For the plants that have been bred and are still bred, including information on the types of changes achieved and the types of radiation used, see the Mutant Variety Database of the International Atomic Energy Agency IAEA: https://mvd.iaea.org/ (31 January 2018). Even though the data are open source there is still no historic or other coherent analysis as to which breeders from which countries produced (and still produce) what kinds of plants and for what particular reasons.
2 Nowadays, the method makes use of artificially produced isotopes, but the first attempts at irradiating plants were conducted with natural substances, so that the expressions "radioactive", "atomic" or "nuclear" are often used interchangeably.
3 Photographs from the 1960s show that plants close to the radiation source were completely burned (Howorth 1960, Appendix, image 28).
4 An example of these data sheets may be found at www.atomicgardening.com/1960/02/21/the-atomic-gardening-society/ (10 January 2018), albeit without archival information.
5 Johnson shows a loose-leaf booklet with a logo of the WAEG with added notes. See www.atomicgardening.com/1960/02/21/the-atomic-gardening-society/ (Stand: 10 January 2018).
6 According to Johnson, Speas was a dental surgeon, but she does not provide source material for the claim. See www.atomicgardening.com/1958/10/20/irradiating-seeds-for-fun-and-profit-c-j-speas-life-magazine-photos/ (Stand: 31 January 2018).

7 In Germany, first tests of the effects of X-rays on bacteria were conducted in 1896 and, only a few years later, first patents for food irradiation were filed in the UK and the USA. According to Zachmann however, no systematic research was established in the early years (see Zachmann 2012, 12). The examples shown here seem to indicate that this is a disputable conclusion.

8 Concerning the history of the institute, see Fengler 2014. Due to the institute's significance the works on irradiation conducted there were probably well known among German-speaking physicists. The general public were interested in medical applications of radioactivity or its various uses as an energy source (see Cirkel-Bartelt 2017).

9 For a transcript of the speech see www.eisenhower.archives.gov/research/online_documents/atoms_for_peace/Binder13.pdf (Stand: 31 January 2018).

10 The appearance of Plutonium fallout in the Earth's atmosphere is currently considered to be the most convincing signal of the beginning of the "Anthropocene" for geological periodization, apart from other aspects like CO_2 concentration or fossilizable biological remains. See www2.le.ac.uk/offices/press/press-releases/2016/august/media-note-anthropocene-working-group-awg.

Bibliography

Anderson, E.G., Longley, A.E., Li, C.H. and Rutherford, K.L. 1949. "Hereditary Effects Produced in Maize by Radiations from the Bikini Atomic Bomb". *Genetics*, 34(6): 639–646.

AÖAW (Archiv der Österreichischen Akademie der Wissenschaften). 1911. "Handschriftliche Mitgliederliste mit Themen". FE-Akten, box 29, Fiche 395.

Balogh, Brian. 1991. *Chain Reaction: Expert Debate and Public Participation in American Commercial Nuclear Power 1945–1975*. Cambridge, MA: Harvard University Press.

Boulding, Kenneth E. 1966. "The Economics of the Coming Spaceship Earth". In *Environmental Quality in a Growing Economy*, edited by H. Jarrett (pp. 3–14). Baltimore, MD: Johns Hopkins University Press.

Boyd, William and Prudham, Scott. 2004. "Manufacturing Green Gold. Industrial Tree Improvement and the Power of Heredity in the Post War United States". In *Industrializing Organisms. Introducing Evolutionary History*, edited by Susan Schrepfer and Philip Scranton (pp. 107–139). New York: Routledge.

Breitwieser, Lukas and Zachmann, Karin. 2017. "Biofakte des Atomzeitalters. Strahlende Entwicklungen in Ghanas Landwirtschaft". *Technikgeschichte*, 84(2): 107–133.

Cirkel-Bartelt, Vanessa. 2017. "Beautiful Destruction. The Aesthetic of Apocalypse in Hans Dominik's Early Science Fiction". *Approaching Religion*, 7(2) (Special Issue): 37–49.

Crookes, William. 1898. "Address of the President before the British Association for the Advancement of Science". *Science*, 8(200): 561–575.

Crutzen, Paul and Stoermer, Eugene. 2000. "The 'Anthropocene'". *Global Change Newsletter*, 41: 17.

Curry, Helen Ann. 2014. "From Garden Biotech to Garage Biotech. Amateur Experimental Biology in Historical Perspective". *British Journal for the History of Science*, 47(3): 539–565.

Curry, Helen Ann. 2016. *Evolution Made to Order*. Chicago, IL: University of Chicago Press.

De Vries, Hugo. 1905. *Species and Varieties, Their Origin by Mutation. Lectures Delivered at the University of California*. Chicago, IL: The Open Court Publishing Company.

Duarte Rodrigues, Ana and Simões, Ana. 2017. "Horticulture in Portugal 1850–1900. The Role of Science and Public Utility in Shaping Knowledge". *Annals of Science*, 74: 192–213.

Falk, Raphael and Lazcano, Antonio. 2012. "The Forgotten Dispute. A.I. Oparin and H.J. Muller on the Origin of Life". *History and Philosophy of the Life Sciences*, 34(3): 373–390.

Fengler, Silke. 2014. *Kerne, Kooperation und Konkurrenz, Kernforschung in Österreich im internationalen Kontext (1900–1950)*. Vienna: Böhlau.

Forgan, Sophie. 2003. "Atoms in Wonderland". *History and Technology*, 19(3): 177–196.

Goldschmidt, Betrand. 1997. "The Origins of the International Atomic Energy Agency". In IAEA, *International Atomic Energy Agency. Personal Reflections*, edited by the IAEA (pp. 1–15). Vienna: The Agency.

Hamblin, Jacob Darwin. 2009. "Let There Be Light [...] And Bread. The United Nations, the Developing World, and Atomic Energy's Green Revolution". *History and Technology*, 25(1): 25–48.

Heim, Susanne. 2003. *Kalorien, Kautschuk, Karrieren*. Göttingen: Wallstein.

Helmstädter, Axel. 2006. "Geschichte der Radiumschwachtherapie. Radioaktivität – die pure Lebenskraft?" *Schweizerische Ärztezeitung*, 87(20): 904–907.

Herrán, Nestor Corbacho. 2008. *Aguas, semillas y radiaciones: el laboratorio de radiactividad de la Universidad de Madrid, 1904–1929*. Madrid: CSIC.

Howorth, Muriel. 1955. *Atom and Eve*. London: New World Publications.

Howorth, Muriel. 1960. *Atomic Gardening for the Layman*. London: New World Publications.

Jevons, William Stanley. 1865. *The Coal Question. An Inquiry Concerning the Progress of the Nation and the Probable Exhaustion of Our Coal-mines*. London: Macmillan.

Johnson, Paige. 2012. "Safeguarding the Atom. The Nuclear Enthusiasm of Muriel Howorth". *The British Journal for the History of Science*, 45(4): 551–571.

Karafyllis, Nicole. 2006. "Biofakte. Grundlagen, Probleme, Perspektiven". *Erwägen, Wissen, Ethik*, 17(4): 547–558.

Kim, Dong-Wong. 2002. *Leadership and Creativity: A History of the Cavendish Laboratory, 1871–1919*. Boston, MA: Archimedes.

Klems, Wolfgang. 1988. *Die unbewältigte Moderne – Geschichte und Kontinuität der Technikkritik*. Frankfurt am Main: Serapion.

Knierzinger, Wolfgang. 2012. "Das Forschungsinstitut Gastein in der Forschunglandschaft des 'Ständestaates' und des 'Dritten Reiches' ". In *Kernforschung in Österreich. Wandlungen eines interdisziplinären Forschungsfeldes 1900–1978*, edited by Silke Fengler and Carola Sachse (pp. 109–129). Vienna: Böhlau.

Kohler, Robert. 1994. *Lords of the Fly: Drosophila Genetics and the Experimental Life*. Chicago, IL: University of Chicago Press.

Lanman, Susan Warren. 2004. " 'For Profit and Pleasure': Peter Henderson and the Commercialization of Horticulture in Nineteenth-century America". In *Industrializing Organisms. Introducing Evolutionary History*, edited by Susan Schrepfer and Philip Scranton (pp. 19–41). New York: Routledge.

Lundqvist, Udda. 2014. "Scandinavian Mutation Research in Barley. A Historical Review". *Hereditas*, 151(6): 123–131.

McNeill, J.R. and Engelke, Peter. 2014. *The Great Acceleration. An Environmental History of the Anthropocene since 1945*. Cambridge, MA, and London: The Belknap Press of Harvard University Press.

McNeill, J.R. and Unger, Corinna R. 2010. "Introduction: The Bigger Picture". In *Environmental Histories of the Cold War*, edited by J.R. McNeill and Corinna R. Unger (pp. 1–20). Cambridge: Cambridge University Press.

Muller, Hermann Joseph. 1927. "Artificial Transmutation of the Gene". *Science*, 66: 84–87.

Perkins, John H. 1997. *Geopolitics and the Green Revolution. Wheat, Genes and the Cold War*. New York and Oxford: Oxford University Press.

Rucht, Dieter. 1995. "The Impact of Anti-nuclear Power Movements in International Comparison". In *Resistance to New Technology. Nuclear Power, Information Technology and Biotechnology*, edited by Martin Bauer (pp. 277–292). Cambridge: Cambridge University Press.

Russell, Edmund. 2004. "The Garden in the Machine. Towards an Evolutionary History". In *Industrializing Organisms. Introducing Evolutionary History*, edited by Susan Schrepfer and Philip Scranton (pp. 1–16). New York: Routledge.

Russell, Edmund. 2010. "Can Organisms Be Technology?" In *The Illusory Boundary, Environment and Technology in History*, edited by Martin Reuss and Stephen H. Cutchcliffe (pp. 249–262). Charlottesville and London: University of Virginia Press.

Shu, Quing Yao and Forster, B.P. 2012. "Plant Mutagenesis in Crop Improvement. Basic Terms and Applications". In *Plant Mutation Breeding and Biotechnology*, edited by Quing Yao Shu, B.P. Forster and Heisuke Nakagawa. s.l.: CABI.

Sievert, Rolf M. 1923. "Secondary Rays in Radium Therapeutics". *Acta Radiologica*, 2(3): 268–300.

Soddy, Frederick. 1903. *Radium. Lecture Delivered at the School of Military Engineering, Chatham on 14th January 1904*. XXIX: 1–17.

Soddy, Frederick. 1921. "Various Newspaper Clippings". Papers and Correspondences, Bodleian Library, Oxford, MS Eng. Misc. box 170, item 16.

Spiller, James. 2004. "The Commercial Fate of Food Irradiation in the United States". *Technology and Culture*, 45(4): 740–763.

Trischler, Helmut. 2018. "The Anthropocene – A Challenge for the History of Science, Technology, and the Environment". *N.T.M. – Journal of the History of Science, Technology, and Medicine*, 24(3): 309–335.

Weart, Spencer. 1988. *Nuclear Fear – A History of Images*. Cambridge, MA: Harvard University Press.

Wieland, Thomas. 2004. *"Wir beherrschen den pflanzlichen Organismus besser". Wissenschaftliche Pflanzenzüchtung in Deutschland, 1889–1945*. München: Deutsches Museum.

Zachmann, Karin. 2012. "Irradiating Fish – Improving Food Chains? Retailers as Mediators in a German Innovation Network (1968–1977)". In *Transformations of Retailing in Europe after 1945*, edited by Ralph Jessen and Lydia Langer (pp. 179–194). Abingdon, Oxon: Routledge.

Zachmann, Karin. 2013. "Risky Rays for an Improved Food Supply? National and Transnational Food Irradiation Research as a Cold War Recipe". *Deutsches Museum Preprint* 7. Munich: Deutsches Museum.

3   Urban utopias and the Anthropocene

Ana Simões and Maria Paula Diogo

Introduction

In this chapter we discuss urban utopias put forward in the early twentieth century, contrasting Portuguese technoscientific with British environmental perspectives. Our comparative discussion builds on parallels between their contexts of emergence and their purported aims. It shows that they are informed by ideological commitments, which convey specific stances on the interplay of technoscientific and natural landscapes in urban settings, providing us with an open window to revisit the human–nature dichotomy, one of the themes central to the Anthropocene debate.

The first set of utopias are the technoscientific utopias for Lisbon presented in the Portuguese journal *Ilustração Portugueza* in 1906: one by the modernist writer and journalist of socialist leanings Fialho de Almeida (1857–1911) and entitled "Monumental Lisbon"; the other by the engineer of Saint Simonian inclination and port expert Melo de Matos (1856–1915) called "Lisbon in the Year 2000". Both authors present Lisbon as a futuristic metropolis deeply shaped by a growing industrial economy, with a new geographical profile anchored in the River Tagus and marked by the rhythm of the new technologies of transport and communications overtaking Europe and the USA, which promised to turn into reality many dreams of the recent past.

The second set is the concept of the garden city put forward by Ebenezer Howard (1850–1928), and which first partially materialized in Letchworth Garden City (Hertfordshire, England), beginning in 1903. In Howard's post-Victorian projects, city and country were integrated through the idealized networked garden cities, echoing influences ranging from industrial utopian socialism and anarchism to Georgism.

Contrary to Howard's environmental economics closely related to the Arts and Crafts Movement, Melo de Matos and Fialho de Almeida describe an industrial ecology that extended the enlightened rationalist tradition of progress grounded on science and technology, and which prefigured many of the urban projects of the futuristic architectural movement.

We argue that their crucial differences depended on the varying urban contexts of their emergence and the ideological visions of their authors. But going beyond their obvious differences, in what relates to the ways in which they envisioned new urban spaces in their relation to both industrial and green sites, we argue that they all shared a view of nature as *techné* which incorporated ingredients that distanced their imagined fabricated nature(s) from primeval/ pristine views in crystal-clear contrast with anything man-made.

Cities, gardens and the Anthropocene

From the Garden of Eden to present-day theories concerning the building of a sustainable world, the history of utopian thinking has been largely the history of human beings' relationship with nature and technology in order to strike a balance between our technologies and the resources they demand from nature. In this chapter we show how urban settings can cast additional light onto one central topic of environmental histories, currently reconvened also in the broader context of the Anthropocene debate.

In the context of the world population growth which followed industrialization, urbanization became the dominant worldwide demographic trend from the nineteenth century onwards, becoming one of the leading twentieth-century phenomena. The number of people living in cities increased from 600 million in 1950 to more than 2 billion in 1986, reaching around 3.9 billion in 2014. It is predicted to reach 6 billion by 2050, according to the *Population Division of the Department of Economic and Social Affairs of the United Nations*. By the mid-twenty-first century, 66 per cent of the world population will live in cities. Of these, in 2014 there were already 28 "mega-cities" worldwide with more than 10 million inhabitants. This number is expected to reach around 40 in 2030, to such an extent that economist Edward Glaeser has gone so far as to name the human species the "urban species"(Glaeser 2011, 1).

The global dimension of urbanization is one of the corner pieces of the debate on the Anthropocene, summoning a whole chorus of voices that appeal to sustainability through myriad solutions based, on the one hand, on technical infrastructures and, on the other, on the "greening" dimension. Ideologically, far from one modernist approach to the city – "the grip of man upon nature [...] a human operation directed against nature" in the words of Le Corbusier (Le Corbusier 1929, xxi) – sustainable cities in the perspective of the "New Urbanism" movement call for a historical affiliation rooted in the garden city movement.[1] Evolving from the original ideas of another utopian modernist, Ebenezer Howard, they share a common view of the role of nature in urban settings, advocating that nature, when more or less tamed, does indeed serve human agency in the city.

For obvious reasons, cities are privileged spaces to reflect on the taming of nature: large technological systems in action – water, electricity, sewage, transportation, housing – rational planning and management of the available space, and everyday lives running to the rhythm of institutional clocks.

At first glance, gardens and parks in industrial cities, neighbourhoods and factories seem to represent symbols of freedom from the urban rational grid, of spiritual communion between human beings and nature – a green splash in the grey city. In addition, since the nineteenth-century park movement, and especially since the work of the American landscape architect Frederick Law Olmsted, gardens and parks, by performing the function of the "lungs" of the city, came to embody a range of human benefits for citizens, mostly in the realm of public health (Olmsted 1870).[2] More recently, many scholars have pointed out the dangerous relations of gardens to power and control. Examples include Leo Marx's examination of American literature representing the interruption of pastoral scenery by industrial technology (Marx 1964), Henri Lefebvre's concept of social production of space, which highlights the dimensions of hegemony and control inherent to any space (Lefebvre 1974), Foucault's Panopticon model that brings surveillance to the heart of modern structures of power (Foucault 1975), Chandra Mukerji's studies on the use of nature to design distinctively national landscapes and create a naturalized political territoriality (Mukerji 1997), and Antoine Picon's analysis of the Parisian park of the Buttes-Chaumon as a display of engineering-based nature (Picon 2010).

However, very little has been written on the epistemological consequences of using nature as a *techné*, thus depriving it from its primal values and transforming it into a concrete, variable and context-dependent object. We argue that this approach to nature as a *techné*, which was particularly striking in the late nineteenth- and early twentieth-century urban utopias mainly concerning the relationship between green and non-green areas, and particularly in those we selected to address in this chapter, is one of the distinctive features of the Anthropocene and is, in fact, at the heart of the debate and criticism of Eco-modernism (Asafu-Adjay *et al.* 2015). It continued to evolve during the twentieth century, reaching a new status on the antipodes of Henry David Thoreau's mid-nineteenth-century conservationist exaltation of nature, in which humankind was presented as part of nature, thus bending to its rules. Thoreau, a leading figure of Transcendentalism,[3] advocated in his book *Walden* (1854) a simple living style in natural surroundings, based on his own experience of living in the woods, in a "Spartan-like way", to use his own words, with the minimum necessary to survive in order to truly understand the meaning of life.

The novel vision of a submissive and human-dependent nature has been characterized by the so-called "technological nature" (i.e. domesticated, artificial or digital representations of nature that fulfil our inborn need of the wild without actually engaging with it). This *lumpennature* (Diogo, Louro and Scarso 2017, 28) is nature contaminated in its very essence, leading to its ineluctable dissolution as a category. As Bill McKibben puts it when addressing nature, human agency and the Anthropocene, "we have deprived nature of its independence, and this is fatal to its meaning. Nature's independence is its meaning – without it there is nothing but us" (McKibben 1989, 58).

In what follows, we use the comparison between Portuguese and British urban utopias to discuss the above-mentioned argument. Guided by selected urban visions, we claim that the question to be asked currently, in the age of the

Anthropocene, is not about the quantity of nature (that is, of natural resources) available to us, but about nature's epistemological quality.

Portuguese and British utopias in the early twentieth century

While Ebenezer Howard's ideal of the garden city, the rise and expansion of the garden city movement, and their materializations in the UK and in other countries have been amply discussed by urban historians,[4] such has not been the case with the less impacting visions of the Portuguese Fialho de Almeida and Melo de Matos, which have never attracted reflections on the cities of tomorrow. To use a blunt down-to-earth metaphor, we propose to compare an elephant with two mice. Unlike most of the scholarship that focuses on canonical urban visions from mainstream countries, we opted to reverse the terms of comparison by contrasting an industrialist utopia stemming from a low industrialized country with an environmentally sensitive utopia from the epitome of industrialization. Therefore, to compare them is a doubly bold enterprise due to their uneven influence and contrasting differences. However, we argue that juxtaposing turn-of-the century urban visions concocted in the country of origin of the Industrial Revolution with those conceived of in a small, peripheral European country struggling to take the first serious steps of industrialization offers an open window to look beyond singularities and touch commonalities ranging across modern trends in urbanization, however disparate.

Despite different contexts of emergence and contrasting authorial ideological commitments, which are analysed in the following section, it is interesting to point out that none of the chosen utopias is futuristic in the sense of dreaming of an unattainable future; on the contrary. all share a pragmatism associated with the drive to propose implementable visions able to actually change the urban landscape for the better. Their authors are not just dreamers but mostly doers, they are not expert city planners but marginal to the emerging profession. But all intertwine proposals of urban renewal with social reform and the improvement of the life conditions of urban populations, especially those afflicting the rampant working class.

Although there is a powerful component of imagination in the two sets of utopias we propose to contrast, they were all deeply moulded by the present. As Michael D. Gordin, Helen Tilley and Gyan Prakash mention in the introduction to the edited volume *Utopias/Dystopias*, utopias are not just "objects of study" per se but also "historically grounded analytic categories" in the sense of futuristic projections which tell us a lot about the present agendas of visionaries; that is, "with which to understand how individuals and groups around the world have interpreted their present tense with an eye into the future" (Gordin, Tilley and Prakash 2010, 3).

The imagined urban landscapes which Fialho de Almeida and Melo de Matos describe are built on their beliefs concerning progress as the outcome of a scientific and technologically driven industrial society; the garden cities of Howard build on a reflection of the drawbacks of industrialized overpopulated cities and the attendant rarified countryside, and propose to square the circle, merging the advantages of both while eradicating their shortcomings.

The city(ies) of Lisbon imagined by Fialho de Almeida and Melo de Matos as well as the networked garden cities of Howard are all strongly designed using technology, and in particular the new transportation systems and infrastructures. Either in their utopias or in other writings, all authors are revealed to be particularly concerned with the well-being of citizens, including the working class, a growing proportion of the city's inhabitants, in what relates to lodging, sanitary conditions and health matters, aligning themselves with the philanthropic and paternalistic tone of utopian industrialism in the late nineteenth and early twentieth centuries.

Under the influence of the model industrial site built in New Lanark, Scotland in the early nineteenth century by Robert Owen, one of the founders of utopian socialism and the cooperative movement, a growing number of industrialists shouldered the responsibility of protecting their workers and providing them with a stimulating and healthy – thus productive – natural (gardens, ponds, vegetable gardens), social (nurseries and schools) and cultural (libraries, theatres, musical bands) environment. By the end of the nineteenth century, the concern for sanitation, health and hygiene became critical to such an extent that urban design incorporated gardens as filtering areas to contain the negative effects of industrial pollution and epidemics, thus expanding to the urban fabric the concept developed in the restricted spaces of factories and working-class quarters. They propelled the change from what Lewis Mumford called "factory camps" (LeGates and Stout 2016, 16), where formerly workers were squeezed into filthy and environmentally polluted areas around their workplaces, to what Helen Chance has recently fittingly dubbed the "factory in a garden" (Chance 2017). Such ideas were stretched to their fullest extent in Howard's networks of garden cities; that is, in what we may appropriately call "the city in the garden" project.

Portuguese utopian cityscapes. Lisbon and its sister city, between reality and fiction

From the mid-nineteenth century onwards, Lisbon underwent a process of modernization and progressive industrialization by expanding towards the north and along the River Tagus, to such an extent that by the 1910s its main urban industrial hubs had been moved to the southern side of the river. Lisbon's population increased by 45 per cent between 1890 and 1910, reaching 436,000 inhabitants in 1911 (França 2005, 63, 68, 72),[5] the first year of the Republican regime.[6] The two techno-scientific utopias for Lisbon discussed in this chapter were published in 1906 in a newly created illustrated magazine, the *Ilustração Portugueza*, associated with the daily journal O *Século*, with a wide readership and closely connected with the republican movement, whose agenda envisioned the creation of a new model of citizenship informed by recent developments in science and technology.

The engineer and port expert Melo de Matos was the editor of a journal called *Construção Moderna* (*Modern Construction*), founded in 1900, the first Portuguese journal to address construction works, and specifically new modern family houses, among which stood dwellings for the working class. His Saint-Simonian predisposition made it a topic especially dear to him, addressed in many of his writings, both from the architectural perspective as well as from the financial one, discussing how workers' associations could find the financial means for their construction (Melo de Matos 1910). Although these matters were not discussed in "Lisbon in the Year 2000", probably because Melo de Matos considered that they would be long solved by then, the urban vision of this port expert understandably focused on the city port as the nerve centre connecting Lisbon to the world. In four successive instalments he depicted the future port, the wharf of Alcântara and its warehouses, the station and the tunnel connecting Lisbon to the south margin of the river.

By 2000 the city port had become an international port of the utmost importance, and an obligatory passage point for foreign navy ships. In its shipyard high-tech ships incorporating innovations by Portuguese experts (engineers, chemists and electricians) and endowed with high-precision instruments were built. The connections of the port to the railway network were established by means of an aerial metropolitan electric system of aerodynamic features, a German invention appropriated by a Portuguese engineer responsible for introducing "exceedingly relevant and practical modifications", endowed with carriages moving periodically "as bright meteors" in the same direction every five minutes (Melo de Matos 1906a, 133). Lisbon had become the central node for commercial transactions with Argentina, Cape Town, Australia and Canada, competing for supremacy with Britain. Aerial metropolitan high-speed transport lines tied the port to its various wharves where commercial transactions were orchestrated by the most sophisticated electric systems of communications, all transactions being supervised from the height of a 350 m-Eiffel-like steel tower, endowed with various elevators and panoramic restaurants.

The metro, together with aircrafts and cars, secured circulation in the principal arteries. Bridges and viaducts completed the circulation network. The architecture of the highly ornamental central station was impressive, as was the commanding position of a huge and very precise clock, simultaneously "the brain and the heart" of the station. Ministries of commerce, industry and communications as well as the main banks of capitalist Lisbon surrounded the station. Among them, special attention was paid to the Agricultural Bank (Caixa Geral Agrícola) whose imposing reliefs symbolized the growing importance of chemistry and meteorology for scientific agriculture. There were plaques paying homage to Liebig, Chaptal, Pasteur and Ferreira Lapa, the Portuguese promoter of scientific agronomy, as well as a marble relief depicting Ceres and modern science "tightly embraced and encircled by laboratory instruments, retorts, mechanical harvesters, farm animals and huge stacks" (Melo de Matos 1906c, 221).

The connection to the industrialized south margin of the River Tagus was described in the first and final instalments. Modern Lisbon, expanding along the north side of the river, blossomed together with its industrialized counterpart, spreading along its south side, where an imposing industrial complex with 25 factories for processing canned fish reshaped the landscape. Various buildings capped with tower chimneys expelling heavy fumes comprised the highly industrialized background to the renewed capital city (Melo de Matos 1906, 133). The underground tunnel connecting the two margins was a feat of technoscientific prowess convening the expertise of foreign and Portuguese scientists, and including highly original technological innovations by Portuguese experts. Underground experiments were immediately made public by phone, telegraph and the press. Portuguese scientists had been raised to the status of national heroes. The trepidation-free aerodynamic tunnel and trains connected Lisbon to Seixal, a locality in the south margin, in just three minutes. Electric lighting and luxurious bathrooms "with all the commodities of civilization" ensured an unforgettable trip (Melo de Matos 1906d, 252).

In the same way as the bright and electric city of Melo de Matos, Fialho de Almeida's vision in "Monumental Lisbon" depicted a cosmopolitan Lisbon run by technology. In addition, it was shaped by aesthetic choices, extending to suburban areas of the city, to gardens and to parks.

Fialho de Almeida attended the Polytechnic School and then the Medical School, having graduated in medicine in 1895, but his professional life centred around journalistic and writing activities, being renowned as a man of letters, not as a physician. He wrote extensively for newspapers and magazines and often published chronicles, narratives, fictional tales and short stories about various aspects of contemporary life, mores and politics, often with a critical, pamphlet-like and satirical bent.

In his late nineteenth-century chronicle *The Cats* (*Os Gatos*, 1889–1894), Fialho de Almeida, together with other outspoken writers of the same group, often criticized the urban choices for the capital. The new palaces looked like "cattle sheds", the new buildings like "dressers", all suited to the "lodgings of idiotic Lisboans". The new Avenida da Liberdade (Liberty Avenue), of Haussmanian inspiration, was full of "hillbilly mansions burping on the avenue, boring and very high".

In "Monumental Lisbon", the expertise of architects in comparison to artisans and amateurs, together with the educational role of the Fine Arts School and the National Society of Fine Arts, demonstrated a specific architectural new urban style, ranging from the monumentality of institutional buildings, bridges and avenues to the creation of a "Portuguese house", all considered fundamental to finding the right balance between modernity, beauty and commodity.

The new building of the Medical School was cited as an example of "a modern building with monumental proportions" (Fialho de Almeida 1906a, 400), and was conceived of as an integral part of a complex of neighbourhoods stretching along the four cardinal points, and representing the medical and health sciences, industry and commerce, culture and leisure time, and finally residential areas, namely the four fundamental vectors for the development of any modern city.

A technological artwork – an imposing bridge – was envisioned to cross the southern part of the imposing Haussmanian-like Avenida da Liberdade, connecting the "medicine hill" on the east, capped by the new grandiose building of the Medical School, to the "sciences hill" on the west, where the Polytechnic School was located, enhanced by a beautiful new gate located at the end of its recently designed botanical garden.

Improvements in the city port were associated with an imposing marginal avenue stretching along the river, often dreamt of by others in the past, and offering citizens open vistas. Finally, the vision of monumental Lisbon, dubbed a "court city filled with rolling orgies, gasps of gas and festivities", was at the same time routinely visited by a "labyrinth of steamers" (Fialho de Almeida 1906b, 498), and was capped by the creation of an "industrial and commercial Lisbon" on the south side of the river. A monumental bridge with two platforms, one for people and the other for trains, connected the two cities of Lisbon, facilitating communication between the two, and of each with the world.

The monumentality of this sister city was derived from its "furnaces and hammers", factories and chimneys emanating heavy smoke exactly like industrial London. To circumvent the problems of industrialization as they materialized in dire living conditions in urban areas in England, the slow but steady progress of the industrialization of Lisbon should avoid concentrating factories and industrial sites close to the river. As such the republican socialist Fialho de Almeida proposed that they all be moved to the new "industrial city" at Tagus' southern margin. Responding also to the filthy and highly populated industrialized London in what relates to workers' living conditions, the new "industrial city" accommodated proletariat neighbourhoods of the "modern hygienic type" (Fialho de Almeida 1906b, 501). They were constructed with appropriate materials, free from contagious diseases, including tuberculosis, with proper ventilation, water and sewage infrastructures, and surrounded by gardens and green spaces where families and workers could spend their leisure time.

Inspired by the traditional rural single-storey houses in the southern regions of Portugal, they were surrounded by small gardens, and were located in rows in large, airy streets with tree-lined sidewalks. Streets radiated from a common large rotunda, which was the heart of the neighbourhood. Amply illuminated and planted with trees, it could accommodate concerts and outdoor activities. It included the public library, the church, a free bathhouse, facilities for children, including a kindergaten and *lactario*, a conference hall, and finally the public school, occupying the richest building of all, and reflecting the importance of moulding the new republican working-class citizen. On the opposite side of the radiating streets a square boulevard marked the neighbourhood's boundary. Planted with trees, it included at its corners playgrounds for children and fields for adults to practise physical exercise and play collective games.

Sketched in considerable detail, but with no accompanying diagrammatic representation, the plan of the new workers' neighbourhoods followed in detail Owen's proposals in order to offer workers natural, social and cultural amenities in a healthy environment conducive to optimal productivity rates. Fialho de Almeida's vision was probably inspired by the "factory in the garden" concept; it also resonated with the recent ideas expounded by Howard in the *Garden Cities of Tomorrow*, where the "factory in the garden" gave way to the more ambitious "city in the garden" ideal.

British utopian cityscapes: the city in the garden

The shared visions of Melo de Matos and Fialho de Almeida contrasted with the ecological utopia of Ebenezer Howard, in which the best of city and countryside intermingled on an unprecedented scale. Put forward in *To-morrow: A Peaceful Path to Real Reform* (1898), they were revised in 1902 as the *Garden Cities of To-morrow*.

Ebenezer Howard was a court stenographer by trade. Born in London, he had experienced since childhood the heavily polluted and overcrowded built-up environment of the modern industrialized city. At the age of 21 he emigrated to the United States and became a farmer at the frontier in Nebraska. He proved to be a disaster at farming, but obtained firsthand knowledge of the Homestead Act of 1862, which enabled the construction of an economy and society of well-off farms and small towns. From 1872 to 1876, he stayed in pre-skyscrapers' Chicago, then known as the Garden City (a possible inspiration for the second title of his book), and began his lifelong career as a shorthand writer. He witnessed the construction of a new garden suburb designed by the landscape architect Olmsted.

Howard returned to Britain in 1876, and soon became involved in discussion circles and political movements concerned with what was then called the "social question". Despite high scholarship, there are still various misconceptions concerning the gist of Howard's main ideas, which stemmed from a variety of sources. It seems unchallenged, though, that he managed to assemble various strands of thought in a highly original synthesis which has influenced urban planning ever since its proposal in the early twentieth century (Hall 2014, 90–148). Urban decentralization, the integration of nature into cities, zoning for different uses, greenbelting and the development of self-contained new towns where communities live on the outskirts of larger central cities became the tenets of Howard's vision and remain the backbone of the tradition of modern city planning.

As already mentioned, Howard was influenced by the ideas of the social reformer and utopian socialist Robert Owen. Other radical theorists influenced him, for example, Peter Kropotkin, the Russian mutualist of anarchist leanings, and Thomas Spence, an advocate of common landownership. Howard referred explicitly to the inspiration of the American Henry George, for whom land was the main source of the wealth of nations. In *Progress and Poverty* (1879), George appropriated seventeenth-century physiocratism which considered land as the sole source of wealth in an economic framework based on incomes and taxes on land, and on new forms of administration of territory, both in terms of space and ownership.[7] Herbert Spencer and John Stuart Mill, both defenders of utilitarianism, provided extra sources of stimulus: from Spencer George borrowed the idea of land nationalization, and from Mill he borrowed the idea of how to do it by means of planned colonization. Unsurprisingly, the agrarian emphasis went hand in hand with an attraction to the anti-industrialist Arts and Crafts Movement. Interestingly, while Howard referred to John Ruskin, one of its major leaders, he never mentioned William Morris and his utopia *News from Nowhere* (1890), which depicts a scenario close to the garden cities.

A final push probably sprang from the cooperative principles of the utopian novelist Edward Bellamy put forward in the futuristic utopia *Looking Backward 2000–1887*. Disliking Bellamy's centralized socialist management based on the collectivization of farms and industries, Howard looked for an alternative that did not subordinate the individual to the group.

Howard proposed the construction of cities of limited size, surrounded by a permanent belt of agricultural lands, the perfect marriage of urban and rural landscapes, administered by citizens and funded by land income. The garden cities formed clusters of "slumless and smokeless cities" (Howard 1898, Diagram no. 7).

To illustrate the ideas on which the concept of the garden city was based, Howard contrasted, in the now famous Three Magnet Diagram, the advantages and disadvantages of town life (first magnet) and country life (second magnet), in order to lay the ground for the balanced town–country model (third magnet), in which people could live in an environment combining the advantages of town and country without any of their major drawbacks.

Howard's clean prose provided compelling arguments for his project. Sharing the ideas of Friedrich Engels about the evils of overcrowding in filthy working-class districts in industrial cities in England, Howard built his argument on the assumption that "men of all parties" agree that urban overcrowding is the horror premise of the urban question. However, people continued to converge upon already overcrowded cities due to the power of attraction of the "town magnet", a combination of social opportunities, places of amusement, plentiful jobs and high wages, which characterized all industrial metropolises. This attraction was unable to counteract the power of the "country magnet", offering close contact with nature, fresh air, an abundant supply of water and bright sunshine, yet in increasingly desolate rural districts. The "town–country magnet" provided Howard's compelling solution to the urban question, since it integrated the best of both worlds, thereby having the potential to attract a new kind of urban community (Howard 1898, 16–19).

Howard also used other diagrams to depict how a hypothetical garden city would be structured. In a concentric ring-shaped diagram he depicted a Central Park where stood important public buildings, surrounded by a Crystal Palace ring of retail stores, encircled in its turn by rings of increasing diameter, where houses and gardens alternated with circular avenues crossed by boulevards radiating from the centre into the periphery. The entire city occupied around 1,000 acres, served a population of around 32,000 people, and was encircled by an agricultural belt of approximately 5,000 acres. When a garden city attained its planned limit, another garden city would start to grow not very far away. Cities were connected to each other by a rapid transportation system, the Inter-Municipal Railway. Extra diagrams explained how various garden cities would be networked to one another via a larger Central City, compounding Howard's vision of the polycentric Social City. Industrial sites and infrastructures formed a belt situated on the outskirts of the garden cities, facilitating transport in and out, and reducing smoke in the central city.

While they were visually compelling, reinforcing the descriptions adduced in the text, explanatory diagrams were not meant to substitute for urban plans and maps. They provided simple but forceful guidelines to be followed in future materializations, which were dependent on the characteristics of the sites to be selected.

Because urban layout went hand in hand with a specific concept of social reform based on landownership in perpetuity by citizens, Howard took pains to detail and provide numerical calculations for his ideas, in order to prove their worth and feasibility. Therefore, he counteracted a third socio-economic way to Victorian capitalism and bureaucratic centralized socialism, based on local management and self-government.

The first materialization of a garden city took place in Letchworth, UK in 1903, in a compromised form, as is often the case, falling short of Howard's initial ideas, especially in what relates to landownership and self-government. Appropriated by the garden city movement, after Howard's death the garden city idea spread and was implemented in continental Europe, in the United States of the New Deal, and in other parts of the world.

Conclusion

In this chapter we advocated that the projected cities of Lisbon put forward by Fialho de Almeida and Melo de Matos be analysed through a lens which goes far beyond the importance of architectural projects in a local perspective. Besides giving us a hint about discussions going on in Europe at the beginning of the twentieth century, we argue that they can be additionally framed in the recent debate on the Anthropocene.

As mentioned above, Melo de Matos was a Saint-Simonian and Fialho de Almeida was a republican socialist. Therefore, for both, progress depended centrally on industrialization, to such an extent that they were not concerned with a change in the regime of the exploration of natural resources, but with the improvement of living conditions of the working class. For both, the mutualist cooperation advocated by Proudhon or the subsequent theories of auto-sufficiency defended by Kropotkin (which influenced Howard) based on small-scale organizations and the imbalance between urban and rural landscapes, that is, between industry and agriculture, were not the utopias they dreamed about. Their utopian visions depended on steam, industry, infrastructures of mobility and global scales.

By contrasting the Lisbon utopias with Howard's garden cities, one is led to tone down the opposition between the "factory in the garden" visions of Almeida and Matos and the "city in the garden" vision of Howard. All utopias encompass a dimension of social reform as part and parcel of urban renewal standing up for egalitarian ideals, but while the Portuguese authors were close to industrial-driven utopian socialism, Howard centred his vision on Georgism and an agrarian and pastoral vision influenced by Proudhon as the main framework for its industrial dimension.

At the same time, the two sets of utopias discussed here converge in how they think about the essence of nature, encapsulated in the gardens and other green spaces in urban settings. Unlike conservationists such as Thoreau, who advocated the inherent value of nature, apart from human usage, Fialho de Almeida's and Melo de Matos' utopias for Lisbon and Howard's projected garden cities, although based on different ideologies, converge in using nature in an operative way – as a *techné*; that is, they appropriate nature to reach specific objectives, and do not just consider it central to individual or collective enjoyable experiences.

This common rationale of thinking about the essence of nature – not its place in the city, and not its changing historical meanings – is, in our perspective, one of the distinctive, albeit evolving features of the Anthropocene. We argue that, from the perspective of the debate on the Anthropocene, the core of the discussion on nature and cities is not whether they oppose each other or are intertwined, but how nature is disempowered and loses its independence from human actions, not in terms of local scales but as part of a global agency. The belief that nature (plants, animals, oceans, forests) is exclusively a resource to be used to solve economic, cultural, physical and mental health problems is one of the main pillars of the "age of humankind", since it justifies the right of human beings to use it in an unrestrained and amoral way, with no concern for the rightness or wrongness of their actions, eliminating any obstacles and supporting a new worldview that affects the way humans think and act. This specific mindset is, in our perspective, closely related to the binomial industrialization-urbanization of the late eighteenth century onwards and marks the beginning of the Anthropocene.

Needless to say, none of the authors discussed was able to predict the scale of use of natural resources which was to take place throughout the twentieth century and well into the twenty-first century. They all shared a somewhat naïve, even religious-like vision of the capacity of natural resources to autoregenerate and continue endlessly to serve human beings' interests.[8] But their utopias offer us the opportunity to better understand the roots of present-day debates and find useful ways to avoid a no-return situation.

Acknowledgements

FCT MCTES – Project PTDC/IVC-HFC/3122/2014 – VISLIS – *Visions of Lisbon – Science, Technology and Medicine (STM) and the Making of a Technoscientific Capital (1870–1940)*.

FCT MCTES – Project PTDC/IVC-HFC/6789/2014 – ANTHROPOLANDS – *Engineering the Anthropocene: The Role of Colonial Science, Technology and Medicine on Changing of the African Landscape.*
FCT MCTES – Project PEst-OE/HIS/UI0286/2014.

Notes

1 The principles of the New Urbanism are laid out in the "Charter of the New Urbanism" published in 1993, but based in former ideas, including those of Howard among many others. The New Urbanism is a visionary planning and design movement that embraces the idea of sustainability and real-estate development based on a small-town scale. It also embraces the importance of natural environment, regional metropolitanism, and the ideals of social justice and participatory democracy. Transcribed in Richard T. LeGates and Frederic Stout, eds, *The City Reader*. London: Routledge Urban Reader Series, 2016, 6th edition (pp. 411–413).
2 Frederick Law Olmsted, "Public Parks and the Enlargement of Towns", American Social Science Association (1870). In Richard T. LeGates and Frederic Stout, eds, *The City Reader*. London: Routledge Urban Reader Series, 2016, 6th edition (pp. 365–370).
3 Transcendentalism is a philosophical movement developed in the United States in the late 1820s and 1830s, anchored in the Transcendental Club (Cambridge, Massachusetts). Based on different Western philosophical traditions and on Hindu texts on philosophy of the mind and spirituality, Transcendentalism's core belief is anchored in the inherent goodness of people and nature, individualism and self-reliance. Transcendentalists were critical of government, organized religion, laws, social institutions and industrialization, adopting progressive stands on women's rights, abolition, reform and education.
4 As a canonical case, there is a vast literature on Ebenezer Howard which is well known to urban historians. See e.g. Robert Beevers, *The Garden City Utopia. A Critical Biography of Ebenezer Howard*. London: Macmillan, 1988); Kermit C. Parsons and David Schuyler, eds, *From Garden City to Green City. The Legacy of Ebenezar Howard*, Baltimore, MD: The Johns Hopkins University Press, 2002; Peter Hall, "The City in the Garden. The Garden-city Solution: London, Paris, Berlin, New York, 1900–1940". In *Cities of Tomorrow. An Intellectual History of Urban Planning and Design since 1880*. Chichester: Wiley-Blackwell, 2014, 4th edition (pp. 90–148).
5 As a way of comparison with other European cities, in 1890, the city's population numbered nearly 300,000, around 10 per cent of the population of Paris, half the population of Madrid, and roughly the same as Bordeaux and Stockholm.
6 The Republican regime overthrew the monarchy in Portugal on 5 October 1910. The new regime aimed not only to enforce a new political vision but also a new model of citizenship based on the values of science and technology and wide access to education.
7 It was to demonstrate Georgian economic principles that the American Elizabeth Magie Phillips invented and patented the game The Landlord's Game, in 1904, which would give way to Monopoly.
8 In this respect, it is interesting to discuss how concepts of "natural capital", coined by E.F. Schumacher in 1973 in his book *Small is Beautiful*, and "ecosystem services", formally introduced in the 1970 report *Study of Critical Environmental Problems* but already in use since the 1940s, reflect these views.

Bibliography

Asafu-Adjay, Johan, Linus Blomqvist, Stewart Brand, *et al*. 2015. *An Ecomodernist Manifesto*. www.ecomodernism.org.

Beevers, Robert. 1988. *The Garden City Utopia. A Critical Biography of Ebenezer* Howard. London: Macmillan.

Bellamy, Edward. 1888. *Looking Backward 2000–1887*. Boston, MA: Ticknor and Company.

Chance, Helen. 2017. *The Factory in a Garden: A History of Corporate Landscapes from the Industrial to the Digital Age*. Manchester: Manchester University Press.

Diogo, Maria Paula, Ivo Louro and Davide Scarso. 2017. "Uncanny Nature. Why the Concept of Anthropocene is Relevant for Historians of Technology". *ICON: Journal of the International Committee for the History of Technology*, 23: 23–31.

Engels, Friedrich. 1892. *The Condition of the Working Class in England in 1844*. British edition.

Fialho de Almeida. 1906a. "Lisboa Monumental. I". *Ilustração Portugueza*, 36: 396–405.

Fialho de Almeida. 1906b. "Lisboa Monumental. II". *Ilustração Portugueza*, 39: 497–509.

Foucault, Michel. 1975. *Surveiller et punir, naissance de la prison*. Paris: Gallimard.

França, Augusto. 2005. *Lisboa: Urbanismo e Arquitectura*. Lisbon: Livros Horizonte.

Glaeser, Edward. 2011. *Triumph of the City: How Our Greatest Invention Makes Us Richer, Smarter, Healthier and Happier*. New York: Penguin.

Gordin, Michael, Helen Tilley, and Gyan Prakash. 2010. "Introduction. Utopia and Dystopia beyond Space and Time". In *Utopia and Dystopia: Conditions of Historical Possibility*, edited by Michael D. Gordin, Helen Tilley and Gyan Prakash (pp. 1–17). Princeton, NJ: Princeton University Press.

Hall, Peter. 2014. *Cities of Tomorrow. An Intellectual History of Urban Planning and Design since 1880* (pp. 90–148). Chichester: Wiley-Blackwell, 4th edition.

Howard, Ebenezer. 1898. *To-morrow: A Peaceful Path to Real Reform*. London: Swan Sonnenschein & Co. (reprinted as *Garden Cities of To-Morrow*. 1902. London: Swan Sonnenschein & Co.).

Le Corbusier. 1929. *The City of Tomorrow and Its Planning*. New York: Payson & Clarke.

Lefebvre, Henry. 1974. "La production de l'espace". *L'Homme et la société*, 31–32: 15–32.

LeGates, Richard T., Frederic Stout, eds. 2016. *The City Reader*. London: Routledge Urban Reader Series, 6th edition.

Marx, Leo. 1964. *The Machine in the Garden: Technology and the Pastoral Ideal in America*. Oxford: Oxford University Press.

McKibben, Bill. 1989. *The End of Nature*. New York: Random House.

Melo de Matos. 1906a. "Lisboa no anno 2000. I. O Porto de Lisboa". *Ilustração Portugueza*, 5: 129–133.

Melo de Matos. 1906b. "Lisboa no anno 2000. II. Os cais de Alcântara e os armazéns de Lisboa". *Ilustração Portugueza*, 6: 188–192.

Melo de Matos. 1906c. "Lisboa no anno 2000. III. A estação de Lisboa-Mar". *Ilustração Portugueza*, 7: 220–223.

Melo de Matos. 1906d. "Lisboa no anno 2000. IV. O tunel para a outra banda". *Ilustração Portugueza*, 8: 249–252.

Melo de Matos. 1910. "Da ação da mutualidade contra as habitações insalubres. Papel do cooperativismo na construção de casas higiénicas e baratas". Lecture presented at the Congresso Nacional da Mutualidade, Lisbon.

Mukerji, Chandra. 1997. *Territorial Ambitions and the Gardens of Versailles*. Cambridge: Cambridge University Press.

Olmsted, Frederick Law. 1870. "Public Parks and the Enlargement of Towns". American Social Science Association. In *The City Reader*, edited by Richard T. LeGates and Frederic Stout (pp. 365–370). London: Routledge Urban Reader Series, 2016 (6th edition).

Parsons, Kermit C. and David Schuyler, eds. 2002. *From Garden City to Green City. The Legacy of Ebenezar Howard*. Baltimore, MD: The Johns Hopkins University Press.

Picon, Antoine. 2010. "Nature et ingénierie: le parc des Buttes-Chaumont". *Romantisme*, 150(4): 35–49.

Simões, Ana. 2019. "From Capital City to Scientific Capital. Science, Technology, and Medicine in Lisbon as Seen through the Press, 1900–1910," in Agusti Nieto-Galan and Oliver Hochadel, eds. *Urban Histories of Science*. London: Routledge, 2019, pp. 141–163.

4 Shaping colonial landscapes in the early twentieth century

Urban planning and health policies in Lourenço Marques

Ana Cristina Roque

Introduction

Over recent years we have seen a growing interest in issues related to urban planning in the Lusophone African countries, emphasizing individual aspects of the relationship between colonialism/urban planning/geographical space in the Portuguese colonial context and the comparison with other colonial contexts in sub-Saharan Africa (e.g. Silva, 2015), as well as the way in which colonial contexts have affected postcolonial urban development (Barros *et al.*, 2014),[1] or the present-day concerns about colonial legacy itself regarding twentieth- to twenty-first-century urban environments and the particularities inherent to its conservation (Mendonça, 2015).

Most of these studies reflect different perspectives and approaches while giving notice of the involvement of several disciplinary areas and the attention paid to precise conditions that, to a greater or lesser extent and according to the specific circumstances of each case, interfere with the organization and structure of the city, its design and functioning and, ultimately, with the relationship between man/place, man/territory and man/nature.

Sharing Paul Jenkins' thesis (Jenkins, 2000), some of these studies assume that the dualist structure of cities like Maputo (formerly Lourenço Marques) or Luanda was the result of colonial policies that made urban planning an instrument of control of colonial power over the territory. However, as underlined by Marat-Mendes and Sampayo (2015: 55), very little is known about the impacts and performance of the principles that guided urban planning in colonial cities, and even less about the options behind those principles and their implications for changing the paradigms of people's lives.

Within this context, and considering colonial policy and economic interests as the main drivers of the city's landscape changes, this chapter discusses the relationship between health/disease/urban growth/environment in Lourenço Marques in the early twentieth century, and how each of these aspects interconnects and interferes with the efficacy of both the Health and Public Office Services determining particular actions, such as the greening of the city, and social policies that, under the heading of Health Policies, served mainly to perform measures of social and racial segregation.

Without specific reference to the binomial health policies/urban planning as underlying the shaping of a colonial landscape, the few studies on the implementation of the Health Services in colonial Mozambique underline the subordination of these services to the objectives of colonialism and its action as one of the key drivers of the colonial system (Shapiro, 1968, Dube, 2009).

The positions of men such as Serrão de Azevedo or Oliveira e Sousa are enough to sustain this hypothesis. Both expressed, openly and explicitly, the social and racial segregation character of the measures proposed by these services, particularly regarding the creation of indigenous districts in the suburbs of Lourenço Marques, in the early twentieth century. Serrão proposed, defended and justified the prohibition of the indigenous to live in the city because

> The way black people live [...] in homes that are true dens, with dirt habits that characterize them, is one of the most powerful elements of urban unhealthy conditions and is, in all respects, a constant danger in the propagation of any epidemics.
>
> (Serrão, 1907: 267)

Sousa endorsed the establishment of indigenous neighbourhoods, since this would allow "an easier monitoring both by the administrative side, as from the health point of view", adding that the ideal would be to concentrate them in a single neighbourhood, as "surveillance will be more effective as lower the number of those neighborhoods" (Sousa, 1908: 269).

Both statements were consistent with the presentation of the Africans as "the problem". The wild nature of the territory and its inhabitants was cited as the main cause of disease and underdevelopment, justifying the urgency to identify, limit, control, neutralize and dominate both (territory and people) to achieve the colonial purposes. To address "this problem" was an essential point of the Portuguese colonial agenda, presupposing a process of domestication and appropriation of nature to support the structuring of a colonial landscape conforming to the Western standards of occupation and development and the concept of progress (Benoist, 2008: 14).

Incorporating the dichotomy man/nature, wild/domesticated (Moscovici, 1974), colonial ideology presented the European as a symbol of civilization, science and technology as opposed to "the other", non-European as uncivilized, superstitious and savage. This "wild nature" was reason enough to legitimize the Portuguese "civilizing mission" as the main driver of the colonial process and, simultaneously, framed the thesis on the success of territorial occupation, as dependent on the possibility of avoiding any contact with the natives that could endanger the health and well-being of the colonists.

In accordance with this ideology, it should be of "great convenience for the health of Lourenço Marques" that Africans and non-African people should be geographically and physically separated, as stated explicitly in the Regulation of the Malaria's Prophylaxis of Lourenço Marques city (Regulamento de Profilaxia anti-Palustre da Cidade de Lourenço Marques), published in 1907.

The association seems obvious, but it does not exclude an interest in a more comprehensive approach, considering both the public health policies relating to urban growth and the scientific development of the late nineteenth century – which are obviously not unique to colonial contexts – and how these policies benefited from this development and were interrelated to become colonial policies, shaping new spaces in accordance with Western interests and values, with a significant impact and changes in the African territorial and human landscape.

Since this subject entails numerous questions that are impossible to address in a few pages, this chapter is limited to a discussion of some of the measures implemented by the Health Services in Lourenço Marques at the turn of the nineteenth century. Among these measures are those that, according to the documents produced by the Health Services Office and the Public Works Office, may be considered to contribute to the construction of a new urban landscape featuring colonial influence marked by intra-urban social and racial segregation.

Lourenço Marques: urban planning arm-in-arm with health policy

In the early twentieth century and consistent with the process of colonial urbanization dating back to the second quarter of the nineteenth century, Lourenço Marques emerged as a promising metropolis in Southern Africa, being the connection with the Witwatersrand area (former Transvaal), in South Africa, was regarded as chiefly responsible for the economic boom driving the development and growth of the city and the port.

The capital of the colony became the stage for analogous situations to those experienced in the Portuguese cites of Lisbon and Oporto when, in the mid-nineteenth century and closely associated with industrialization, these two urban centres became attractive locations for all those seeking job opportunities and better living conditions (Cosme, 2006). Similarly, Lourenço Marques became a melting pot of people and cultures, where people of all backgrounds, crafts and nationalities mingled, highlighting the need for structural changes and precise measures to address the problems arising from the increasing number of inhabitants largely resulting from the growing rural–urban migration, and, simultaneously, offering the opportunity for major changes in the territorial and human landscape design of the colony.

As noted by Diogo and Navarro (2018), the creation of building infrastructures was the basis for the concept of anthropogenic landscapes in the colonial territories in Africa, oriented and structured in the perspective of European models and interests. The African colonial landscape should embody and reflect the concepts of civilization, development and comfort according to European standards, since only Europeans could be the drivers of progress and pioneers of multiple changes. Consequently, reflecting both the Portuguese colonial imaginary and the specific concerns related to the European process of industrialization, housing, health, safety and the healthiness of factories and workshops, hygienic food preparation and water quality became special concerns and key areas of critical intervention.

In line with these concerns, from the mid-nineteenth century onwards, new government agencies were created or reorganized[2] while specific legislation was discussed and published to define and regulate different measures to improve people's living and working environments, with an emphasis on the sanitary conditions of housing, workshops and industrial establishments.[3]

Sanitary concerns clearly gained importance as part of the health care policy in Portugal with immediate impact on the Portuguese colonies, as testified by the publication, in 1860, of the General Regulation of the Health Service of the Overseas Provinces (Regulamento Geral do Serviço de Saúde das Províncias Ultramarinas) and other specific regulations, such as the Maritime Health Regulation (Regulamento Geral de Sanidade Marítima) of 1897.

With particular reference to Mozambique and in spite of a first regulation in 1844 and the creation of a Healthcare Company in January 1875, the launch of a significant number of sanitary measures, namely those concerning the drainage of the swamp or the publication of the Regulation of the Health Services of the Colony of Mozambique (Regulamento dos Serviços de Saúde da Colónia de Moçambique) in 1878,[4] closely follows the creation of the first hospital of Lourenço Marques in 1878, the administrative reforms of the colony and the increasing development of the city which, in 1898, became the capital of the colony.

From about the last quarter of the nineteenth century and clearly after becoming the capital, Lourenço Marques witnessed an economic, urban and demographic boom which imposed a wide range of policy instruments and measures regarding health, sanitation and public works with visible results in the city's urban landscape and its social and economic geography. Although these measures were often insufficient or even inadequate, they did not fail to demonstrate the political and economic will to transform the unhealthy village of the early nineteenth century into the cosmopolitan metropolis of the first decades of the twentieth century according to a logic of progress explicit in a model of development based on the exploitation of the Africans and expressed, with clear visual impact, in the dualistic structure of city/suburb, centre/periphery and white city/black townships.

Avoided because of the endemic fevers but appreciated for its excellent harbour and easy connection to the mines of the Rand, Lourenço Marques grew and developed quickly regardless of the lack of infrastructure and threats to public health. None of these contingencies has, in fact, compromised the expansion of the city, but they did make clear the need for government intervention in the process of planning and organizing the city as well as providing basic structures and healthier conditions to respond to the increase in the population. In this context, the arrival of the first Public Works team in 1877 should be seen as proof of this intention to organize and manage the urban and business area of the city, but also to improve existing structures and to build the infrastructures still needed to ensure effective possession and control of the territory in accordance with the principles resulting from the Berlin Conference (1884–1885).[5]

Draining the swamp and greening the city

Headed by Joaquim José Machado, the Portuguese military engineer responsible for the study of the first connection of the Lourenço Marques–Pretoria railway line in 1879 (Machado, 1886) and later governor general of Mozambique (1890–1891), this first Public Works team focused its action on two fundamental axes with immediate impact on the city, namely the port and the people's daily life and, ultimately, on the structuring and consolidation of Portuguese colonial power in Mozambique.

On the one hand was continuation of the works to eliminate the swamp area whose landfill and drying had already been foreseen and included in the annual budget of the Municipality for 1875 (Meneses, 1883: I–56). This included the construction of a dike for water containment, the opening up of a ditch collector and the launching of the construction of a first sewage system with direct discharge into the bay, enabling the drainage of a large part of the swamp waters, providing the necessary conditions for the elimination of mosquitoes, and thus contributing to building a healthier city. According to Longle, about 114 hectares of land needed to be cleaned; an insignificant area when compared with the expected growth of the city at the turn of the century (Longle, 1887: 16).

On the other hand, the investment in reinforcing transport and communication infrastructures (railroads, highways, ports) aimed to improve and secure international connections, along with the drafting of the city's urban plan based on a reticular layout, projecting the expansion of the city to the higher and ventilated areas and foreseeing the paving and greening of the streets and of the drained swamp, as well as the definition of specific areas for the settlement of the native population which, together, would contribute to the "advancement of the health issue" in Lourenço Marques.

Accordingly, and in line with previous proposals that had drawn attention to the problem of the marshlands (Barros, 1845), afforestation projects were also welcomed. Greening the city was as much a requirement from a health point of view as a key element in the city's beautification and well-being of its inhabitants but, above all, it was an expression of progress and civilization through mastering nature and transforming the city into a domesticated landscape.

For some areas, experts from the Public Works Office, unaware of the potential of the local flora but in line with the recent works on afforestation processes in Portugal (Ribeiro e Delgado, 1868, Júnior, 1879) suggested combining the drainage works with the planting of large, power-absorbing trees such as *Eucalyptus globulus* or *Paulownia imperialis* for marshes, *Casuarina tenuissima* planting suitable for sand-dune areas and for protection against wind and dust (Longle, 1887), or different types of trees for shading sidewalks, streets and avenues, while other areas were assigned by the Municipality to the city's residents to be transformed into gardens or allocated to institutions specifically created for the greening of the city.

The *Municipal Garden Vasco da Gama*, nowadays known as Tunduru Gardens, fits precisely into the latter situation as it emerged from the initiative of the

Horticulture and Floriculture Society of Lourenço Marques (*Sociedade de Horticultura e Floricultura* of Lourenço Marques), founded in 1885, to "improve the hygienic conditions of the city and its surroundings by means of afforestation"[6] and to provide all inhabitants of and visitors to the city with "the pleasure of enjoying the garden and all the wooded land".[7]

Having as its main objective the afforestation of the swamp, which surrounded the city from the north and southwest, and the creation of a garden, the *Sociedade* was supported by the City Council and the government of the colony (Sousa, 1951: 59) and, despite nearly 50 years' delay, responded to the royal recommendations of 1838[8] in order to create a botanical garden in Mozambique that should contain "the most interesting African plants and will serve to acclimatize those of other parts of the globe" (Ribeiro, 1893: 398).

In fact, the garden was initially used as a nursery for botanical garden species, including *Eucalyptus* whose seeds were provided by the Public Works Office (Sousa, 1951: 59–60), and served as an acclimation garden and test field for plant species, particularly for those used in swampy, sandy or dry areas,[9] namely different species of *Eucalyptus* and *Casuarina*.

The initial area comprised about 13 hectares and was half drift sand/ half malarial swamp (Sim, 1919 after Sousa, 1951: 59). Despite the interest and support of the colonial government and the Ministry of Overseas (Longle, 1887), Sousa emphasizes that the transformation of this barren and unhealthy place was only possible because of the human will and persistence of the colonists involved in the making of a garden, which thus appeared as a powerful symbol of civilization itself (Ritvo, 1992).

In fact, from November 1885 until the end of the century, the garden represented a stage in the process of the domestication of nature consistent with the colonial concept of a civilizing mission. The clearing of the land of the wild and sickly marsh to give way to a garden complex was a kind of metaphor of the colonial project itself, expressing the powerful role of European science and technology in both environmental changes and the building of modern colonial empires (Malhi, 2017).

The environmental impact of the works was huge. The area was fenced, the wetlands dried and drained, existing watercourses were recovered, a large lake was built, the first plantings were made, and nurseries of various species were created responding to different seasonal climatic conditions. The original idea of planting various species of eucalyptus was expanded with the possibility of testing other species, whose seeds were sent from the metropolis or obtained in neighbouring countries – Natal and Cape colonies – enabling the development of new nurseries of plants more appropriate to the regional climatic conditions.[10]

In 1887, a small botanical garden had already been properly delimited and fenced, anticipating what would become, in the early twentieth century, the city's Municipal Garden. Opened to the public in 1900, the garden of the former *Sociedade* came to attest to what was expected to be an acclimatization garden organized to meet the scientific, pedagogical and recreational leisure requirements of the inhabitants, but in accordance with the modern concept of a botanical garden (Ritvo, 1992) and the characteristics of each deployment space (Ribeiro, 1893: 409). The result was the creation of a landscape garden with a clear anthropic imprint reflecting man's power over nature, and the ability to shape it to their interests. As Thomas Sim stated in a report in 1909 on the malarial swamp that had cost Lourenço Marques its bad reputation and many thousands of lives in the past had been transformed into a beautiful Municipal Garden for the pleasure, recreation and health of the city's inhabitants and visitors (Sim, 1919 after Sousa, 1951: 59).

In late January 1887 and following an agreement with the Government of the District and the Municipality,[11] the garden nurseries were already in position to provide the Public Works Office with more than 3,200 handmade palm leaf baskets (*capachos*) with eucalyptus seedlings ready for field planting in marshy and flooded areas. Towards the end of the century the number of nurseries had increased, a small eucalyptus grove at the edge of the swamp was already a perfumed barrier to protect the city from the emanation of unhealthy gases and mosquito clouds, and the organization of the garden as a leisure space enabled its opening to the public in 1900.

Consistent with the metropolitan arborization projects of the nineteenth century and the creation of public green garden areas in Lisbon, such as *Monsanto*, *Campo Grande* or the *Jardim da Estrela* (Rodrigues, 2017), the Garden of the *Sociedade* would become an oasis of lush greenery in the middle of the city, a recreation area for repose and retreat, and a meeting point for the residents, a symbol of the European colonizer's ability to tame the wild and unhealthy African nature and mould the African landscape to colonial needs and standards.

Under the responsibility of Thomas Honey, the garden was developed between 1907 and 1920 with the planting and acclimatization of new species, mostly exotic and ornamental (Sousa, 1951: 63–66), as well as the annexation of nearby areas, namely Vila Joia, property of Gerard Pott, the Dutch Consul in Lourenço Marques. Therefore, by the mid-twentieth century, the initial area had been much enlarged and benefited from the new landfills and lawns which allowed the elimination of the swamp's residual stagnant waters and transformed the entire area into a "wooded place and lawn, full of warm shadows and varied colors" (Sousa, 1951: 59)[12] according to an eclectic plan that, combining classic and romantic designs, gave the garden its own identity.

Greening the city was thus an important component of the overall sanitation project to create conditions for the growth and development of the city, promote a healthier environment and contribute to improving the lives of its inhabitants. According to Longle (1887: 58), the process should encompasse drainage where it was possible to harness the waters, planting eucalyptus whenever the soils allowed it, and landfilling all lowlands and unsuitable areas for any type of crop, and presupposed the existence of previous geomorphological studies as well as knowledge on the ethology of the plant species to be used according to the specific characteristics of each type of soil. Science and technology were used to testify to the knowledge and efficient management of the territory in the creation of a new city landscape and, as noted by Diogo and Navarro (2018: 113), "technical objects and systems became the visible face of modernity" and the architype of civilization and progress.

Therefore, other than the recovery and reconversion of the marshy areas, the grid proposed in the 1880s encompassed new uptown residential areas, such as Ponta Vermelha, the recovery of downtown business centres and the rehabilitation of the surrounding neglected areas of Mahé and Maxaquene, as well as the creation of gardens, the construction of a new hospital and other medical and sanitary facilities, but all under the logic of the creation of a "white city" and assuming that the indigenous population would be confined to predefined areas. These areas had already been identified as *Bairro Indígena* (Indigenous neighbourhood) in the grid presented in 1887 by António de Araújo, in his *Projecto de ampliação da cidade de Lourenço Marques* (Figure 4.1), foreshadowing the idea of the Africans as a source of disease and cause of unhealthy environments as later advocated by Serrão de Azevedo in 1907.

Araújo was a member of Machado's team and seems to have been primarily responsible for drafting the first urban plan of Lourenço Marques. His plan would be the first in a series of plans culminating in the 1950s with the General Plan for the Urban Development of Lourenço Marques, authored by João António de Aguiar, extending the urban grid outlined in the nineteenth century and persisting in the same logic of social and racial segregation, as was clear from the first urban projects.

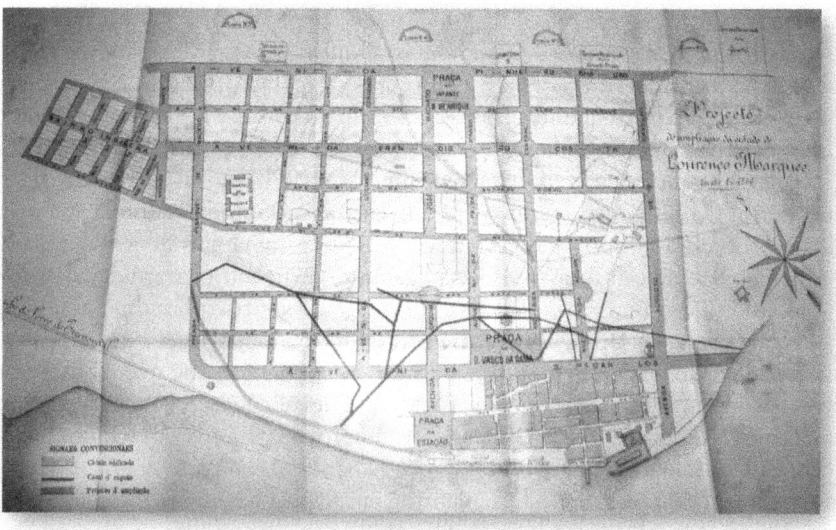

Figure 4.1 Project for expanding the city of Lourenço Marques, 1887.
Credit: Commons.

However, up until then, the city grew on the sidelines of any predefined urban or social plan, regardless of existing or planned sanitary conditions, and, in 1899, according to the Head of the Health Services, Lourenço Marques, "having developed materially to a point that only eyewitnesses can understand, had completely neglected the health of their inhabitants" (Barreiros, 1899).

Lourenço Marques had a population of about 5,000[13] inhabitants at the time, but this number quickly increased. Based on the statistics published in April 1906, Freire de Andrade refers to a population of about 9,000 inhabitants, of which 30 per cent were indigenous (Andrade, 1950: 28).

Although most people had to share their daily lives in the city's business centre around the harbour and the bay, social status and racial divisions determined severe restrictions and variations in living conditions, predicting social and economic imbalances and inequalities that would be reflected in the geography of the city, in the way the city would be organized as living and cultural space, and ultimately in the emergence of different urban and cultural landscapes.

The indigenous neighbourhood designed by Araújo was not built at the time, but the city grew in parallel with the asymmetrical economic, social and racial relations between colonizer and colonized. On the one hand, the rich colonial neighbourhoods included in the new urban plan covered by sanitation projects and landscape gardens; on the other hand, the peripheral neighbourhoods, where poor natives and foreigners, whether European or not, survived on hunger wages and under miserable conditions classified as "hygienic barbarism" (Barreiros, 1899), and the downtown areas where the concentration of people of all origins and crafts called for the building of infrastructures to cope with urban growth.

Sanitation and infrastructures

At the end of the nineteenth century, the sanitary situation of Lourenço Marques was not very different from that experienced in European cities at the time (Abellán, 2017), but the specific geo-climatic characteristics of the region, combined with the growing number of inhabitants living in deplorable circumstances, contributed to deteriorating health conditions.

The city's poor sanitary situation resulted in serious and worrying health risks. Despite the new municipal regulations and interdictions,[14] the city was still a mix of architectural buildings, wooden houses and barracks of wood and zinc plate, while wild bushes continued to grow on the banks of the city's busiest streets and "its decomposing mixed with organic waste and exposed to the high temperature and the seasonal torrential rains" (Barreiros, 1899) was responsible for turning the unpaved sandy streets into mud puddles where, during the rainy season, mosquitoes proliferated, and people were forced to walk around up to their ankles in water.

The sewage system was rudimentary and limited to certain parts of the city, garbage collection was almost non-existent, and many people persisted in opening up ditches in their yards to drain away all kinds of animal and human waste fostering foci of infection, attracting bugs and parasites, and contaminating soil and groundwater. The water was expensive and of poor quality. The city's running water, not suitable for drinking but a luxury available only to very few, was considered "vitiated of extraneous substances (and a) good fluid for microbial cultures".[15] Commodities such as water, food and lodging were also tremendously expensive; there was no marketplace, no public graveyard and no slaughterhouse. Much of the food products were sold door to door by the producers or their employees, house backyards were used as burial grounds, and animals were slaughtered out of doors in yards and courtyards under the most anti-hygienic conditions and without any sanitary control.

Medical assistance was rudimentary, and hospital facilities lacked conditions and space necessary to fight disease and treat the increasing number of patients. In the first decades of the twentieth century the hospital, temporarily installed in Ponta Vermelha, was still without a cistern and the water used there, coming from the Umbelusi River, was unsuitable for all types of use. It could not be used in the kitchen, for washing or remedies,[16] patients could not bathe, and even the sterile water, piped from Mocímboa for medical purposes, was littered with mosquito larvae.[17]

The combination of these conditions provided an unhealthy atmosphere, conducive to the proliferation of endemic and epidemic diseases which often originated in the hospital facilities. Thus, the regular and seasonal occurrences of marsh fevers, respiratory and gastrointestinal diseases joined the incidences of dysentery, tuberculosis, syphilis, leprosy, tapeworm and even scabies. Together they were enough to declare "unsatisfactory" the health conditions of the city and district, regularly mentioned in all the Health Services' reports. However, these were not the only scourges afflicting the inhabitants of Lourenço Marques.

Between 1890 and 1909, the city (and the district) was the stage for several epidemics (Roque, 2016: 170–171), together with outbreaks of tuberculosis (1901 and 1902) and malaria (1907 and 1909), which affected the entire population, regardless of race, gender or age. The virulence and the endemic nature of many of these diseases triggered reactions and raised public and institutional concerns, the outcome of which was the approval and adoption of several health and hygienic sanitary measures.[18] Despite the ideas about the role of reforestation as a driver of climate change and of promoting a healthier environment (Ribeiro e Delgado, 1868), these new measures did not encompass any investment in the use of specific flora species for health purposes. Landscape gardening and street-shading trees were not foreseen for areas other than those already existing or planned. These new measures were strictly related to preventive and prophylaxis actions, such as large-scale vaccination campaigns (Azevedo, 1908), and a greater regularity of house inspections, considered essential for the eradication of what were regarded as major scourges, namely smallpox and plague affecting mostly Africans, and malaria affecting mainly Europeans.

Since smallpox and plague were infectious diseases forcing quarantine and isolation of potential patients, a complex of basic hygiene and sanitary measures was created to combat both, and, as early as 1898, a sanitary police unit was created to supervise the application of these measures through home inspections. Tasked with the identification and removal of infected patients to isolation pavilions and quarantine those possibly infected, this unit was ordered to evacuate the patients and burn down their houses, and to carry out the mandatory vaccination of all inhabitants in case of smallpox, propose sanitary improvements to the area if possible or, ultimately, to burn down the entire village, forcing the remaining inhabitants to leave and settle elsewhere, in areas predefined by the colonial authorities. Thus, measures and procedures were framed by specific regulations which established a set of responsibilities, shared by the Health Services, the local administration and the municipality, to be performed in concert and according to each situation (Ribeiro, 1917: 258–259, 266–269, 278).

These actions were considered crucial to eradicate the origin of these two diseases and had significant environmental, social and economic impact in both the areas where they were applied, which once free of natives could be reclassified and "domesticated", and in the areas defined to relocate those who were forced to move, which, in most cases, did not have the conditions or resources to support any new settlement. Since this relocation process resulted from the very dynamics of the city's expansion process and the need to ensure the safety and well-being of its residents, these actions became a key element in the process of the construction of the colonial landscape.

In addition, since it also entailed the existence of space for the treatment of patients infected with smallpox and plague or in quarantine, it was necessary to construct specific facilities,[19] and, while waiting, to find a temporary solution to keep these patients isolated and limit the chances of contagion. Most of the patients were housed in existing sanatoriums, such as the *lazarettos* of Magude, or in the nearby islands of Elefantes and Xefina,[20] sharing space and medical care with patients with different pathologies. The exception, for some of those infected with smallpox, was the temporary use of a house in a secluded location near the garden of the *Sociedade*,[21] suggesting the idea of healing gardens and therapeutic landscapes (Gerlach-Spriggs *et al.*, 1998) to improve patients' health and provide a healthier environment as generalized in Europe for the treatment of tuberculosis (McCarthy, 2001).

Most of the proposed measures focused on sanitary practices calling for hygiene principles and space classifications related to their main functionality – workspaces, warehouses, housing, leisure areas – and, consequently, requiring precise conditions or needs. Their effectiveness presumed the clear separation between each one of these spaces and institutional and individual commitment in the implementation of the measures in every workplace, house, public place or common area.

Hence, regarding workplaces, it was mandatory to empty the stores and warehouses, which had to be cleaned with sulphurous acid, the walls whitewashed and the buildings properly ventilated before storing grain or other products, and ensuring that warehouses were not used as dormitories for employees. As for households, owners and residents were responsible for whitewashing both interior and exterior walls and demolishing all the uninhabitable wooden houses and tents, as well as cleaning cisterns, tanks and water reservoirs, terraces, yards and courtyards, burning all vegetal and animal waste and eliminating all equipment, products and causes of unhealthy situations, including ponds and standing water, that could endanger public health.

In addition, the Health Service Office had to check the conditions of each dwelling according to the latest scientific theories on the "space scientifically considered necessary to each individual to breathe", censoring all those without the necessary requirements or enforcing improvements, and the maximum number of people per room.[22] However, despite the scientific basis of the measures proposed and the foreseen sanctions,[23] most of these procedures could barely be met by the poorest African, Asian or European families living in crowded dwellings in poor conditions, in the suburbs of the city, and acted as determinants in the design of the new city's urban and human landscape and its concomitant social and economic geography.

As most of these measures were considered crucial to improving general living conditions and public health, they were subsequently incorporated into the Regulation of the Malaria's Prophylaxis (1907) as essential for promoting a healthy environment adverse to mosquito breeding.[24] Thus, at the beginning of the twentieth century, disease-fighting strategies were based on a combination of principles and procedures, which benefited from the investment in strengthening preventive measures (vaccination campaigns, health visits, isolation pavilions, etc.) as much as in a significant number of public and sanitary works (drying of swamps, sewage and waste removal systems, water treatment and improvement of the water distribution network, creation of gardens, paving of streets, new hospital facilities, etc.) which, theoretically, would benefit all citizens.

Colonial policy: Health Services and public works as drivers of social inequalities

Both the Health Services and the Public Works Office seemed to agree when it came to articulating public health policies and urban plans to mould the city's landscape to colonial interests, the indigenous and non-European population (Indians, Asians, etc.) being the preferred target of most preventive measures and the white population the beneficiary of the city's sanitation improvements, as predicted in Araújo's plan (Figure 4.1).

In fact, Araújo's plan anticipates the concept of apartheid as a structuring axis of colonial policy and implies that the Public Works Office and the Health Services are important key players in the colonial system (Shapiro, 1968, Dube, 2009). The identification and demarcation of an indigenous neighbourhood, outlined as a small module adaptable to future expansion needs (Tostões and Bonito, 2015: 43), attests to the idea of segregation and social inequality underlying the structure of the city's plan and its paramount importance to sustain the complex of health measures and sanitary procedures proposed to improve environmental conditions and combat major diseases.

The combination of sanitation works, vaccination campaigns and new methods of diagnostics and treatments, reflecting the significant improvements in tropical medicine and public health, was an important contribution towards improving the city's healthy environment. However, most actions performed by the Health Services in Lourenço Marques, rather than acting effectively in the prophylaxis of diseases, relegated them to the city's periphery, where no sanitary intervention or housing and greening programme was planned.

In the early twentieth century, the expanding suburbs of Lourenço Marques replicated the conditions eradicated in the city, contributing to accentuate the dichotomy between the white city (architectural buildings, water distribution system, large paved avenues shaded by trees, gardens, hospital facilities, etc.) and the black suburbs (adobe or wattle and daub houses, no running water, unpaved tracks, kitchen gardens, no medical care, etc.) (Vales, 2014: 18), attesting to the shaping of different cultural, social and economic landscapes and making clear the differences between urban and suburban areas as well as between the different social and ethnic groups.

Lourenço Marques developed from this visible contrast in the city's plan, as drafted by Araújo in 1887, and later inscribed in the health regulations. As defended by Azevedo, particular places should be chosen in the periphery:

> to establish a settlement for indigenous people and another for Asians. In these settlements, which should conform to a plan previously defined, only the construction of rudimentary houses of reed, wood and zinc should be allowed, while concrete buildings were forbidden.
>
> (Azevedo, 1907)

These houses needed to replicate the same structure as the ones destroyed and interdicted in the city, this option being justified for reasons of public health and the need to defend the well-being of white residents. Within the city they were a potential source of infection and contagion, threatening the non-indigenous population living in concrete buildings: hence the importance of demolishing poor housing and housing estates and the interdiction of rebuilding them in the city. But, on the periphery, where the danger of contagion of Europeans was reduced, they needed to be the prototype to use, as it allowed rapid demolition, "without great losses", whenever there was urgent need (Azevedo, 1907).

In this context, since African natives were considered primarily responsible for sanitation problems and the major diseases affecting the city, epidemics could be easily eradicated from the city as soon as the natives were prevented from living there while forcing them to live in predefined peripheral districts, making it easier to control any epidemics.

Under these conditions and in case of epidemics, the existence of these neighbourhoods would facilitate their easy and rapid destruction and prevent contagion with other zones.In the absence of any urban plan and sanitary infrastructure network to guarantee rapid reconstruction, the remaining materials of the previous demolition were recycled where possible, and the "reproduction" of the neighbourhood facilitated in another location to where the native population would be forced to move. Extra investments were thus considered a waste of money and energy that could be used to improve the city's healthier environment and the well-being of its inhabitants.

Segregation, therefore, emerged as a key element in creating a city free from epidemics. In the early twentieth century, health policies and urban and sanitation projects were instrumental in shaping a new city landscape compatible with Western concepts of progress and civilization but working as vehicles of an apartheid-based colonial policy.

Final considerations

During the first decade of the twentieth century, Lourenço Marques was a test bed for multiple control strategies to combat major endemic and epidemic disease. This involved the implementation of specific preventive regulations and significant public and sanitation works. These procedures were in line with those applied in Europe (Abellán, 2017) and benefited from the latest scientific developments in medical and public health (Bynum, 1994; Cartwright, 1977; Worboys, 2011) applied to tropical diseases. However, in relation to the indigenous population, there was no improvement in living conditions or medical care.

Blamed for their lack of hygiene, promiscuity and "bad" habits, the indigenous population was considered the source of all environmental and sanitary problems affecting the city, and therefore an obstacle to a healthy and harmonious urban development (Roque, 2015). Consequently, for hygienic reasons the indigenous population was forbidden to live in the city and was relegated to the suburbs. There, medical care and sanitation were summed up in the incursions of the health authorities for compulsory vaccination, and prophylaxis was limited to finding possible sources of infection among the natives who could have escaped from the city's health authorities but, in a demarcated neighbourhood, could be easily controlled.[25]

Despite the existence of proposals for the creation of indigenous neighbourhoods on the outskirts of the city, these settlements grew out of various plans drawn up during the first decades of the twentieth century, being the immediate and visible result of the strategy of social and racial segregation imposed by the colonial system. Indeed, while its construction did not take place, "a wide range of policy instruments, laws, and institutions were used to force domestic mobility [...] and thwart people's ability to settle wherever they wanted" (Turok, 2012: 4). Thus poor families, indigenous or otherwise, were pushed to the city's periphery for economic reasons and as a result of racial discrimination.

Benefiting from "the extensive transnational flow of planning ideas, principles, methods and tools that marked the entire colonial period among the metropolises in Europe and across the colonial boundaries in Africa" (Silva, 2015: 15), Lourenço Marques developed according to a pattern of growth which, reflecting the progressive appropriation of space and its accompanying needs, conformed to Marcel Poëte's idea of the city as a response/mode of human adaptation to needs arising from the gradual conquest of nature (Poëte, 2000: 5).

Araújo's urban plan realized this idea. Lourenço Marques was destined to be a modern, civilized and healthy city, projected to be a great metropolis in Southern Africa and probably one of the most well-designed cities in the world (Noronha, 1895). According to Noronha, Araújo's plan was the matrix for the transformation of the "insignificant and tortuous village into a large, open, ventilated and beautiful city occupying an area of 100 hectares" (Noronha, 1895: 99), in which landscaped gardens were a guarantee of health.

The promotion of a healthy environment through the greening of the city was inseparable from the idea of progress and expressed the power to tame and shape wild nature to the interests of civilization. As Noronha commented, Araujo's plan incorporated "human adaptation" by proposing concrete forms of appropriation of different spaces. However, its reference model applied only to the "white city" and not to the indigenous neighbourhoods, and kept open the option for the adoption of different patterns of urban organization. The answer to the "needs arising from the conquest of nature" was heterogeneous, expressed differently according to the dichotomy of city/suburbs, centre/periphery, and was a form of racial segregation in urban development responding primarily to the need for cheap migrant labour to support colonial economic interests (Turok, 2012).

Araújo's modern, civilized and healthy city expanded in a cosmopolitan manner. But his plan, as a powerful instrument of colonial policy, privileged the well-being of the white colonial elites to the detriment of improving the conditions in which most of the population lived and worked. In this context, the use of health and sanitary issues to change the city landscape was a way of asserting the superiority of the colonial system and its ability to reproduce and impose the Western models of progress and comfort in colonial territories.

Acknowledgement

Project FCT UID/HIS/04311/2013. CH-ULisboa – *Empires, Nature, Science and Environnement*.

Notes

1 See also the long list of Paul Jenkins' works listed at www.academia.edu/10355313/ Paul_Jenkins_CV.
2 *Ministério das Obras Públicas, Comércio e Indústria* (1852); *Junta Consultiva de Saúde Pública* (1868); *Direção Geral de Saúde e Beneficência Pública* (1899).
3 *Regulamento de Saúde Pública* (1837); *Código Penal* (1852); *Regulamento sobre a salubridade das instalações industriais* (1855); *Regulamento sobre a salubridade das edificações urbanas* (1902); *Reforma da organização dos serviços de saúde, higiene beneficência pública* (1899).

4 Regulation elaborated and approved according to the law of 28 May 1896, and the decree of 2 December 1869.
5 *Revista de Obras Públicas e Minas*, no. 353–354, 1899, p. 382.
6 Estatutos da *Sociedade de Arboricultura e Floricultura de Lourenço Marques*. Cap. 1, 1885.
7 Estatutos da *Sociedade de Arboricultura e Floricultura de Lourenço Marques*. Cap. 1, 1885.
8 *Portaria de 19 de julho de 1838*. In Ribeiro 1893: 398.
9 *Direção Geral de Agricultura, Ofício nº 217 de 3 de Novembro de 1886*, AHU, ACL SEMU, DGU, Moç.Cx.1389/1L.
10 Parecer do Ministério do Ultramar, 28 July 1887, AHU (cota ACL_SEMU_DGU_Mç_Cx.1389/1L).
11 Following a proposal made by the Society to the Government of the District and the Municipality, the garden was committed to supply up to 4,000 eucalyptus seedlings in *capachos*, ready to be planted, with a height of not less than 0.25 cm, receiving for each seedling 225 reais (Longle, 1887: 31).
12 For a general overview of the history of the garden and its changes throughout time, see the references to the Projecto de Reabilitação do Jardim Tunduro (2012) in Mendonça (2015).
13 The statistics, published in 1897, give Lourenço Marques a population of 4,902 inhabitants: 2,243 Europeans, 913 Asiatics and 1,746 Africans (Delagoa Bay Directory, 1899: 22).
14 "Postura sobre Edificações". *Boletim Municipal de Lourenço Marques* (15), 26 June 1897: 2.
15 *Relatório anual da Província de Moçambique – Anno de 1903*.
16 *Hospital Provisório da Ponta Vermelha – Boletim sanitário de Janeiro* (23 January 1918).
17 *Hospital Provisório da Ponta Vermelha – Boletim sanitário de Janeiro* (27 January 1918).
18 For example, construction of the slaughterhouse (1890), the public cemetery (1891), the municipal market (1903) and the opening of the new hospital of Lourenço Marques (1911); *Health Regulation Project for the City* (1893); reorganization of the *Maritime Health Regulation* (1901); regulation measures to combat smallpox with specific instructions for its application (1901); publication and dissemination of the information on the procedures to combat the pestilence (1901 and 1902); the city's *Regulations for Malaria Prophylaxis* (1907) and the discussion project on the *Medical Assistance to the Indigenous People* (1907).
19 The first sanatorium for patients infected with smallpox was ready to open eight years after the first smallpox cases were diagnosed in the city (*Boletim Sanitário de Lourenço Marques – Janeiro de 1898*) and though it was also used in other instances it became a key player in the fight against this epidemic. Regarding the plague, patients were forcibly isolated in *lazarettos* built outside the city. However, the threat of an outbreak of a "South African plague", in 1901, led the Board of Health to take additional preventive sanitary measures extended to the whole city and therefore with greater impact on public health.
20 *Boletim Sanitário de Lourenço Marques – Janeiro de 1898*.
21 According to the *Boletim Sanitário de Lourenço Marques – Maio de 1898*, this house was probably part of the housing complex formerly belonging to the Dutch consul. Most houses in the garden were part of this complex and over the years were associated with different services, from health to assistance services. In 1951, after having served as a primary school, one of these houses became the headquarters of the public assistance service (Sousa, 1951: 59), but there is no information on the specific benefits of such an environment in the case of smallpox.
22 *Boletim Sanitário de Moçambique – 1901*.
23 Imprisonment and interdiction of inhabiting the city were among the penalties foreseen for non-compliance or violation of the sanitary measures.

24 Malaria required additional measures of protection, such as the use of mosquito nets around beds, windows and doors, the elimination of curtains inside houses and climbing plants outside, but also "the use of oil to cover stagnant waters" (petrolatum) was difficult to eliminate to avoid the concentration and action of mosquitoes. And, although it was not a contagious disease, isolation was recommended to keep patients out of the disease transmission cycle.

25 *Boletim Sanitário da Província de Moçambique*, 1912.

Bibliography

Abellán, J. (2017), "Water Supply and Sanitation Services in Modern Europe: Developments in 19th–20th Centuries". Conference paper presented at the *XII Congreso Internacional de la Asociación Española de Historia Económia*. Universidad de Salamanca, 6–9 September 2017. Available at www.academia.edu/34533153/Water_supply_and_sanitation_services_in_modern_Europe_developments_in_19th-20th_centuries (accessed January 2018).

Andrade, A.A. Freire de (1950), Relatório sobre Moçambique (1906–1910), 2nd edition, 2 Vols, Lourenço Marques.

Araújo, A.J. de (1887), *Projecto de Ampliação da Cidade de Lourenço Marques*. PEUMM, Vol. I, 2008, p. 22.

Azevedo, J. de O. Serrão de (1907b), *Relatório do Serviço de Saúde de 1907*. AHU, 1514 DGU 5ª Repartição. Moçambique (1908) Serviço de Saúde.

Azevedo, J. de O. Serrão de (1908), *Relatório do Serviço de Saúde da Província de Moçambique Lourenço Marques, 30 de Setembro de 1908*. AHU, 1514 DGU 5ª Repartição. Moçambique (1908) Serviço de Saúde.

Barreiros, A.G. (1899), *Lourenço Marques – Boletim Sanitário do mês de Agosto de 1899*. AHU, 1514 DGU 5ª Repartição. Moçambique (1897) Serviço de Saúde.

Barros, C.P., Chivangue A. and Samagaio A. (2014), "Urban Dynamics in Maputo, Mozambique". *Cities*, 36, pp. 74–82.

Barros, F.J. de (1845), *Ofício no. 196/1845 de Filipe José de Barros, Cirurgião mór de Quelimane e Rios de Senna, Moçambique, 25 de Dezembro de 1845*. AHU 1507 DGU 5ª Repartição. Moçambique (1845–1868). Serviço de Saúde – Ofícios dos Empregados da Saúde.

Benoist, A. (2008), "A Brief History of the Idea of Progress". *The Occidental Quarterly*, 8(1), pp. 7–16.

Boletim Municipal de Lourenço Marques (15), 26 June 1897.

Boletim Sanitário da Província de Moçambique elaborado pelo chefe do Serviço de Saúde de Lourenço Marques relativo ao ano de 1912. AHU, 1527 – UM DGC 8ª repartição. Moçambique, 1911–1912, Serviço de Saúde.

Boletim Sanitário de Moçambique relativo ao ano de 1901. AHU, 1514 DGU 5ª Repartição. Moçambique (1898) Serviço de Saúde.

Boletim Sanitário de Lourenço Marques relativo ao ano de 1898. AHU, 1514 DGU 5ª Repartição. Moçambique (1898) Serviço de Saúde.

Bynum, W.F. (1994), *Science and the Practice of Medicine in the Nineteenth Century*. Cambridge: Cambridge University Press.

Cartwright, F.F. (1977), *A Social History of Medicine*. Harlow: Longman.

Cosme, J. (2006). "As Preocupações Higio-Sanitárias em Portugal (2ª metade do século XIX e princípio do XX)". *Revista da Faculdade de Letras HISTÓRIA*, III Série (7): 181–195.

90 *Ana Cristina Roque*

Diogo, M.P. and Navarro, Bruno J. (2018), "Re-designing Africa: Railways and Globalization in the Era of the New Imperialism". In D. Pretel and L. Camprubí (eds) *Technology and Globalisation: Networks of Experts in World History*. Palgrave Studies in Economic Studies, pp. 105–128. Basingstoke: Palgrave.

Dube, F. (2009), *Colonialism, Cross-border Movements, and Epidemiology: A History of Public Health in the Manica Region of Central Mozambique and Eastern Zimbabwe and the African Response (1890–1980)*. PhD thesis, University of Iowa. Available at http://citeseerx.ist.psu.edu/viewdoc/download;jsessionid=AD03357CB67738BF18B1F0454E5D6C59?doi=10.1.1.466.5688&rep=rep1&type=pdf (accessed November 2017).

Gerlach-Spriggs, N., Kaufman, R. and Warner, S.B. Jr. (1998), *Restorative Gardens: The Healing Landscape*. New Haven, CT, and London: Yale University Press.

Hospital Provisório da Ponta Vermelha – Boletim sanitário relativo ao ano de 1918. AHU, 1514 DGU 5ª Repartição. Moçambique (1918) Serviço de Saúde.

Jenkins, P. (2000), "City Profile: Maputo". *Cities*, 17(3), pp. 207–218.

Júnior, O. (1871), *Almanach do Horticultor: Breve Notícia sobre o Eucalypts Globulus e a utilidade da sua cultura em Portugal*. Porto.

Longle, A. (1887), *Do saneamento e alargamento da villa de Lourenço Marques, Pelo Conductor d'Obras Publicas da Provincia de Moçambique, Armando Longle S.S.G.L.* Lisbon: Typografia de A. da Costa Braga.

Machado, J.J. (1886), *De Lourenço Marques a Pretória. Comunicações à Sociedade de Geographia de Lisboa, acerca de alguns assuntos relativos ao Transvaal, ao Caminho de Ferro de Lourenço Marques e à Província de Moçambique* (sessões de 9 e 16 de novembro e 2 e 14 de dezembro de 1885). Lisbon: Imprensa Nacional.

Malhi, Y. (2017), "The Concept of Anthropocene". *Annual Review of Environment and Resources*, 42, pp. 77–104.

Marat-Mendes, T. and Sampayo, M.F. de (2015), The *Plano de Urbanização da Cidade de Luanda* by Étienne de Groërand and David Moreira da Silva (1941–1943). In Silva, C. Nunes (ed.) *Urban Planning in Lusophone African Countries*, pp. 57–78. Aldershot: Ashgate.

McCarthy, O.R. (2001), "The Key to the Sanatoria". *Journal of the Royal Society of Medicine*, 94(8), pp. 413–417.

Mendonça, L.Â. Franco de (2015), *Conservação da Arquitetura e do Ambiente Urbano Modernos: a Baixa de Maputo*. PhD thesis, Universidade de Coimbra. Available at www.academia.edu/30313681/Conservação_da_arquitetura_e_do_ambiente_urbano_modernos_a_Baixa_de_Maputo (accessed November 2017).

Meneses, J.G. de Carvalho e (1883), *Relatório do Governador da Província de Moçambique*, Vol. I (1872–1875), Lourenço Marques.

Moscovici, S. (1974), *Hommes domestiques, hommes sauvages*. Paris.

Noronha, E. de (1895), *O Districto de Lourenço Marques e a África do Sul*. Lisbon: Imprensa Nacional.

Parecer do Ministério do Ultramar, 28 July 1887, AHU, ACL_SEMU_DGU_Mç_Cx. 1389/1L.

Poëte, M. (2000), *Introduction à l'Urbanisme*. Paris: Sens & Tonka.

Regulamento de Profilaxia anti-Palustre da Cidade de Lourenço Marques (1907), Lourenço Marques.

Relatório anual da Província de Moçambique relativo ao ano de Anno de 1903 (1903). Lourenço Marques.

Revista de Obras Públicas e Minas (1899), Ano XXX, Tomo XXX, nos 353–534, May/June.

Ribeiro, C. e Delgado (1868), *Relatório acerca da arborização geral do país*. Lisbon: Tipografia da Academia Rela das Ciências.

Ribeiro, J.S. (1893), *História dos estabelecimentos scientíficos, literários e artisticos de Portugal nos sucessivos reinados da monarquia*. Lisbon: Typographia da Academia Real das Sciencias.

Ribeiro, S. (ed.) (1917), *Anuário de Moçambique*. Lourenço Marques: Imprensa Nacional.

Ritvo, H. (1992), "At the Edge of the Garden: Nature and Domestication in Eighteenth and Nineteenth-Century Britain". Symposium: An English Arcadia: Landscape and Architecture in Britain and America). *Huntington Library Quarterly*, 55(3), pp. 363–378.

Rodrigues, A.D. (2017), *Horticultura para todos*. Lisbon: BNP.

Roque, A.C. (2015), "The Growth of Lourenço Marques at the Turn of the Nineteenth Century". In Carlos Nunes da Silva (ed.) *Urban Planning in Lusophone African Countries*, pp. 101–110. Aldershot: Ashgate.

Roque, A.C. (2016), "Doenças endémicas e epidémicas em Lourenço marques no início do século XX. Processos de controlo versus desenvolvimento urbano". *Anais do Instituto de Higiene e Medicina Tropical*, 16, pp. 167–174.

Shapiro, M.F. (1968), *Medicine in the Service of Colonialism. Medical Care in Portuguese Africa (1885–1974)*. PhD thesis, University of California, Los Angeles.

Silva, C.N. da (2015), "Colonial Urban Planning in Lusophone African Countries: A Comparison with Other Colonial Planning Cultures". In Carlos Nunes da Silva (ed.) *Urban Planning in Lusophone African Countries*, pp. 7–28. Aldershot: Ashgate.

Sim, T. (1919), *Flowering Plants and Shrubs for the Use in South Africa*. Cape Town.

Sousa, A. de F. Gomes e (1951), "O Jardim Municipal Vasco da Gama de Lourenço Marques. Notícia comemorativa do seu cinquentenário". *Boletim da Sociedade de Estudos da Colónia de Moçambique*, 21(68), pp. 58–80.

Sousa, Oliveira e (1908), "Ata da Reunião da CMS da Cidade de Lourenço Marques. 1908". In J. de O. Serrão de Azevedo (1907), *Relatório do Serviço de Saúde de Moçambique relativo ao ano de 1907*, p. 267. AHU, 1514 DGU 5ª Repartição. Moçambique (1908) Serviço de Saúde.

The Delagoa Bay Directory for 1899. A.W. Bayley & Company, Lourenço Marques and Barbeton.

Tostões, A. and Bonito, J. (2015), "Empire, Image and Power During the Estado Novo Period: Colonial Urban Planning in Angola and Mozambique". In Carlos Nunes da Silva (ed.) *Urban Planning in Lusophone African Countries*, pp. 43–54. Aldershot: Ashgate.

Turok, I. (2012), *Urbanisation and Development in South Africa: Economic Imperatives, Spatial Distortions and Strategic Responses*. Urbanization and Emerging Population Issues – Working Paper 8. London: International Institute for Environment and Development.

Vales, T.C. (2014), *De Lourenço Marques à Maputo: genèse et formation d'une ville. Architecture, aménagement de l'espace*. PhD thesis, Université de Grenoble. Available at www.theses.fr/2014GRENH012 (accessed January 2018).

Worboys, M. (2011), "Practice and the Science of Medicine in the Nineteenth Century". *Isis*, 102, pp. 109–115.

Part II
Gardening the Anthropocene

5 From *Pairidaeza* to Planet Garden

The *homo-gardinus* against
desertification

Ana Duarte Rodrigues

Introduction

The Anthropocene refers to a geo-historical period through which human beings have had a significant global impact on the Earth's geology and eco-systems. The definition itself does not hold a negative connotation, but the changes caused to the Earth's ecosystems led to the conclusion that man became the biggest threat to life on Earth, to such an extent that he may even be responsible for the extinction of the human species, as most pessimists will argue (Zylinska 2014). In 1991, Clive Ponting published a landmark work enti-tled *A Green History of the World*, whose subtitle "The Environment and the Collapse of Great Civilizations" has been used as a warning against eminent Ecocide.[1]

Sing C. Chew's most recent book, *World Ecological Degradation: Accumula-tion, Urbanization, and Deforestation 3000 BC–AD 2000* (2001), is more or less its follow-up, and Chew concludes that the relationship between culture and nature has always been exploitative and, consequently, the history of civiliza-tions is also the history of ecological degradation and crisis. However, others argue that some pre-industrial cultures have lived in ecological balance for thou-sands of years[2] and it has only been over the past 65 years that the relationship between culture and nature has deteriorated, depriving the ecosystems of their balance and impoverishing the societies living there. The ruling form of capit-alism leading to the planet resources' exhaustion, industrialization without limits, and war and lack of democracy in many parts of the world lie behind climate changes and the acceleration of the planet's destruction. The Anthro-pocene, dominated by the *homo-economicus*,[3] became the biggest threat to life on Earth, to such an extent that it may even lead to a true Ecocide. To over-come the ruling form of capitalism and to counterbalance a world dominated by the *homo-economicus*, I propose that one should embrace the attitude of the *homo-gardinus*. To offset a world dominated by the *homo-economicus*, in the sense that human actions are "atomistic and self-interested" (Ng and Tseng 2008, 266), I propose that one should embrace the thoughts and actions of the *homo-gardinus* as a metaphor based on gardeners from all times who, by taking care of others – the plants – have helped build a Paradise on Earth.

I will use the ancient roots of the Persian garden – *pairidaeza* – to tackle planet Earth in such a way that I will approach the local garden as a metaphor for thinking about "Planet Garden". Following the perception of planet Earth from the Moon on 20 July 1969, Gilles Clément developed the concept of "Planet Garden" as the Earth was suddenly perceived as a closed, finite garden which requires wise maintenance. These considerations should have very concrete consequences for the way in which we act. If the Earth is Planet Garden we should all be gardeners, experts on water-wise landscaping, in seedlings, pruning, grafting and other horticultural practices.

In this chapter, I will show how the attitude of the *homo-gardinus* was able to fight desertification by applying the praxis of a gardener on a larger scale. Taking as a point of departure the inspiring story of Jean Giono's *The Man Who Planted Trees*, I highlight four examples of in fighting desertification: the action of the single-handed reforestation in the Sahel, two national projects in the USA and China, and finally, a continental-scale project in Africa.

Fighting desertification is currently a major challenge for the *homo-gardinus*. According to data from the Rio de Janeiro Earth Summit in 1992 (reinforced in Rio de Janeiro Earth Summit 2012), by then desertification affected 70 per cent of all drylands, threatening 3.6 billion hectares (equivalent to a quarter of the world's land area), and, consequently, affecting almost one-sixth of the world population (Pastemak and Schlissel 2011, 20). Sixty per cent of the world's food aid is distributed among the populations of the African arid lands (Sasson 1990). Desertification is defined by the United Nations Convention to Combat Desertification (UNCCD) as "land degradation in arid, semi-arid and dry sub-humid areas resulting from various factors, including climatic variations and human activities" (United Nations Environmental Programme (UNEP) 1994). Moreover, the general perception of desertification is associated with the degradation of non-desertic regions, mainly after what occurred in the Sahel region since the 1970s (Heshmeti and Squires 2013, 5).

In present times, most experts find cultural factors to be the basis of desertification, and many are related to fast development that disregards sustainability factors. For example, Pasternak and Schlissel stress that desertification is caused by man through activities such as "overgrazing, over-cultivation, cutting too many trees and by unwise irrigation" (Pasternak and Schlissel 2011, 33). Although difficult, it is possible to reverse desertification (Pasternak and Schlissel 2011, 33).

Stemming from a global perspective, Anton Imeson states: "desertification and land degradation can be stopped, and large areas restored of their quality" (Imeson 2012, xi). On a global scale, much of it relies on the impact of the United Nations Convention to Combat Climate Change and on the European Soil Conservation Policy's actions. Imeson states that it is possible to combat desertification both locally and globally, as the processes of land degradation are similar everywhere (Imeson 2012, xii). According to him, desertification is above all an eco-cultural process characterized by human-induced land degradation (Imeson 2012, 5).

This framework does not include the formation of deserts such as the Sahara, which result from biophysical processes and are decoupled from human processes. In the Sahara's case, a change in the Earth's orbit some 5,500 years ago led to a change in the solar radiation on the Earth's surface and the Savanna-like cover of the Sahara became a desert (Geist 2017).

I argue that gardening can reverse desertification to such an extent that forests, parks, agroforestry and even vegetable gardens can be created on degraded land if and when the gardens' and landscapes' epistemological framework is at stake. Although deserts cannot be transformed into productive land as the reasons why they are deserts are grounded on purely biophysical processes caused by a subtle change in the Earth's orbit, desertification can be prevented (Geist 2017, ch. 1). Desertification, as the process of turning into desert-like conditions regions that were formerly green, can be reversed, as I show with the following case studies.

From *Pairidaeza* to Planet Garden

Pairidaeza is an ancient Persian word used to define an enclosed park, a garden (Messervy 2007, 97). Etymologically, the very root of the word literally means "around" (*Pairi*), and *daeza* is the origin of the words *dezh* or *diza* which in modern Persian mean "enclosure". *Daeza* also has another root in the Indo-Iranian verb "dhaizh", which originally meant "to construct out of earth", but also means "to construct out of nature", as etymologically "nature" means the existing totality.[4] This highlights the idea of a garden as a place where man and nature meet: man as the one who builds, in and with earth/nature. Moreover, the idea of building for the Persians was firmly bound to the ultimate goal of creation, which, according to Mazdaean-Zoroastrian ideology, embodies a certain "happiness for making".

This is the primordial notion of the garden, distinct from *natura, natura* (wild nature) because man enclosed it and defined its boundaries. Consequently, the *pairidaeza* was not only a closed place but also a protected and controlled one – a place where order prevailed. The Farsi word *pairidaeza* is at the root of the Greek word *paradeisos* and the Latin word *paradisus*, establishing a lasting connection between garden and Paradise. The word "Paradise", as the very image of a celestial garden, ultimately entered most European languages and Western culture, as Christian faiths associate Paradise with the Garden of Eden – a state of harmony between man and nature prior to the fall from grace. All depictions of the Garden of Eden consistently emphasize its exquisite beauty, its abundance of water, and the profusion of plants and/or animals. Therefore, and stemming from both their etymological pathway and cultural layers, gardens are where man and nature meet to produce a bountiful place. A place of joy and beauty.

The Western garden has its origin in this *Pairidaeza* and that is why the image of Paradise for us could only be the one of a garden, inasmuch as a real garden is always a piece of Paradise. Stemming from its beginning, the garden is a privileged space where culture meets nature, and where art works with nature. Making a garden is synonymous with intervention in nature, but, unlike the practice of agriculture, involves the making of form and meaning in the sense of art. Moreover, since creating a garden is composing with living things, the gardener knows from experience that on the one hand one can control nature to a certain extent and, on the other, one cannot determine all aspects of the plants' development or end.

In my opinion, this knowledge and the awareness of its limits is the advantage of the *homo-gardinus* in facing the major challenges of our times: increasing natural disasters due to climate change, increasing demands for water, food and energy, and overexploitation leading to desertification, are pushing the planet to its limits. Therefore, due to the planet's current circumstances (i.e. suffering from the negative effects of the Anthropocene), I propose to go back to the concept of the "Planetary Garden".

When Gilles Clément formulated the notion of the "Planetary Garden" some 45 years ago, inspired by the vision of planet Earth as seen from the Moon, he grasped the Earth as a finite and fragile entity that had to be cherished and cared for (Clément 2015). From one day to the next, on 20 July 1969, planet Earth was suddenly perceived from the outside as something finite, with limited resources, and its horizon became henceforth the whole universe. Planet Earth became the closed, finite garden, the *pairidaeza* which requires wise maintenance and protection from the wild, no more the *natura, natura*, but the universe that represented the unknown and the wild.

Our perspective shifted from the *pairidaeza* to "Planet Garden" essentially due to a technical event. Equipped with this global perspective of the territory, we had to renew the conceptual frameworks in which this new world vision would be fully grasped. The garden, taken to be local up until then, became the appropriate concept to think about the global, the planet.

These considerations should have very concrete effects in the way in which we act. Due to the diverse environmental problems of current times globally categorized under the umbrella of climate change, I argue that the comparison of the Earth with a garden should lead us to action and that only if humans embed gardeners' attitude in their actions can we reverse the destructive path in which we have placed the planet and ourselves. Therefore, if the Earth is Planet Garden, then we should all be gardeners, experts on water-wise landscaping, in seedlings, pruning, grafting and other horticultural practices.

Taking as a point of departure the inspiriting story of Jean Giono's *The Man Who Planted Trees*, I will highlight some examples of how we can create a positive impact of man's action over nature to fight desertification and contribute to avoid a true Ecocide. Therefore, I took stock of some cases of afforestation where either a sole man or 20 governments pooled their efforts to act as the *homo-gardinus*. Moreover, and following Bruno Latour, efforts to join nature and culture in an actor network built through interaction, "non-human" actors such as trees should be considered as part of this story (Latour 2004).

Jean Giono's *The Man Who Planted Trees*

The French writer Jean Giono (1885–1970) wrote a short text with the goal of making the act of planting trees enjoyable. It was published for the first time in

English in 1953. The novel *The Man Who Planted Trees* became the author's most acknowledged work. Its long success – several editions, multiple translations, its adaptation to a short animated film by Frédéric Back in 1987, and its theatrical adaptation by Richard Medrington in 2006 – is due to the timelessness of its message. As the leading actor of the story, Elzéard Bouffier transformed the landscape to such an extent that it benefited climatic conditions, soil capacity, the breeding of flocks of sheep and the social relationships among the people of his village, he became a hero of the present ecological movement (Scruton 2014).

In *The Man Who Planted Trees*, Jean Giono tells the story of a shepherd who lived in an isolated and desolate valley of the Alps. In 1913, the narrator went on a lone hiking trip to this region to savour the wilderness and found a desert-like valley, without any trees or any trace of civilization. Only wild lavender grew there. He was looking for water when he met the middle-aged shepherd Bouffier who took him to a natural spring he knew. At this point the narrator became puzzled with this lonely man, who had found peace and stimulus in transforming the landscape after becoming a widower. The shepherd decided to cultivate 300 trees a day, spending his nights choosing the best seeds to plant the following day. He made holes in the soil with his curling pole and dropped into the holes seeds he had brought from distant areas of the forest. Besides his dedication and patience, day after day he had the help of livestock movement. Life in the nearest village was also fading as people could not socialize with each other due to climatic adversity. Giono tailored the peasants' personalities to match the environmental conditions and they are described as mean and selfish.

The narrator participated in World War I. This fact is introduced by Giono to enrich his thoughts about nature, war and technology. Nature and the countryside lifestyle are the purest of man's idylls, while the modern technological society is imbued with the schism of war (Lawrence 1972, 590). In 1920, disappointment takes the narrator to the Alps again. However, he no longer finds the desolate valley:

> The oaks of 1910 were then ten years old and taller than either of us. It was an impressive spectacle. I was literally speechless and, as he did not talk, we spent the whole day walking in silence through his forest. In three sections, it measured eleven kilometers in length and three kilometers at its greatest width. When you remembered that all this had sprung from the hands and the soul of this one man, without technical resources, you understood that humans could be effectual as God in other realms than that of destruction.
>
> (Giono 1953, 8)

Bouffier represents Giono's statement of the liberty of an individual to do good in all its simplicity, as opposed to the rising dictatorships in the first half of the twentieth century (Lawrence 1972).

The narrator was overwhelmed with surprise. He surrendered to the beauty and peacefulness of the place and would return every year to visit Bouffier. The valley had been transformed by Bouffier into a kind of Garden of Eden. Moreover, since the trees had grown, the windy front that prevented people from leaving their houses to socialize with each other also disappeared. Hence, by establishing a correspondence between the very cold, dry and hostile landscape and the bad feelings of people dominated by isolation and selfishness, the story of the planting of these trees became an example of how the creation of this forest changed the climate that, in turn, promoted socialization. People became cheerful and generous. It is a story of how afforestation can change people. It is a story of hope.

In fact, Jean Giono's story highlights man's ability to completely transform the valley landscape to such an extent that his forest was even misconstrued as being "natural". Giono introduces an episode to highlight how this single forest could be mistaken for a natural landscape. Therefore, Bouffier's forest received official protection in 1935, as the authorities were astonished at the accelerated growth and beauty of this valley. This protected the forest from charcoal burning. And the mystery of the new forest was revealed by the narrator to a friend who served on the Forest Committee. They opted to remain silent and decided to leave Bouffier just the way he was, as he had "discovered a wonderful way to be happy" (Giono 1953, 10).

Bouffier is an inspiration, and in the Anthropocene era we need Bouffiers more than ever. Drawing upon this story, I searched for real Bouffiers, for men who have changed the landscape, who have transformed desolate places into "green sites", and, consequently, have improved the life of the people around them. Although not unique, the African Yacouba Sawadogo is the most astonishing man "who has planted trees".

Yacouba Sawadogo: *The Man Who Stopped the Desert*

Yacouba Sawadogo became well known through a film made by Mark Dodd, the writer, producer and director of *The Man Who Stopped the Desert*.[5] Sawadogo stands out as the man who fights desertification in Burkina Faso, a country in the Sahel region, landlocked between six countries without a sea coast and, even by African standards, one of the poorest countries.

In Burkina Faso, especially following severe droughts in the 1970s, the fight against the spread of the desert was undertaken by non-governmental organizations (NGOs), government agencies and development projects, and significant tree-planting operations took place, including the National Village Forestry Programme and the campaign "8000 Villages, 8000 Forests" (Sawadogo *et al.* 2001, 35).

Much has been planted, but few plants or trees have grown. None was as successful as Yacouba Sawadogo's actions. In 2001, he was already regarded as "an outstanding farmer innovator" in the village of Gourga, northwest Burkina Faso (Sawadogo *et al.* 2001, 35). In the futuristic vision of Paul Hanley entitled *Eleven* (2014), which evokes the 11 billion people populating planet Earth by the end of the twenty-first century, Yacouba Sawadogo is presented as one of the most "inspiring stories of ecological restoration and poverty reduction [that] can be found around the world" (Hanley 2014, 160).

Moreover, the environmental scientist Dorkas Kaiser defines him as a "scientist" since he never gives up experimenting. Following David Edgerton's conceptual framework of "technology-in-use" (Edgerton 2006), Sawadogo's reinvention and use of Zaï holes stands as a paradigmatic case in which slightly improved traditional horticultural practices become more efficient than modern technology or new inventions.

Following the severe droughts in the Sahel in the 1970s, most farmers decided to leave the region. However, Sawadogo decided to stay on his father's land. Together with another innovative farmer, Mathieu Ouédraogo, they transformed the landscape through decades of empirical research. They used ancestral horticultural practices in the region, called the "cordons pierreux" and the "zaï holes", as a starting point. Both techniques seek to take maximum advantage of the scarce rainfall levels that occur in the region. In the first case, they used lined, fist-sized stones perpendicular to the flow of rainwater to facilitate the distribution of water and silt along the process. In the second case, they dug holes in the form of semicircles that were also used to catch water. This deserves further explanation, as it became the cornerstone for Sawadogo's success.

Why is the Zaï technique reinvented by Sawadogo so special? The first step is to dig holes in crusted land (a process called the "Zippelle"), with a diameter of 20 to 40 cm and a depth of 10 to 20 cm. Pits are dug during the dry season (from November to May) and the number of Zaï pits per hectare varies from 12,000 to 25,000 (Kaboré and Reij 2004). The excavated earth is ridged around the semicircle to improve the water retention capacity of the pit. Sawadogo introduced some changes in this ancestral technique and he not only enlarged the size of the hole but he also incorporated manure in the dry season, so that the Zaï holes became the ideal horticultural technique to grow trees in the region. Following the first rains, the matter is covered with a thin layer of soil and the seeds are placed in the middle of the pit. This solution stimulates a chain reaction. Dried and compacted soil makes it difficult for water to soak in and helps roots develop. Therefore, the introduction of manure into the Zaï holes became a crucial step for landscape reformation, as manure attracts termites that are renowned for being efficient excavators to break up underground tunnels. As the soil becomes less compact, seeds begin to sprout. As plants grow and their roots go deeper into the soil, structure is created and humidity increases, to such an extent that after three decades the ecosystem and landscape changed – and a forest of 20 hectares was created, including more than 60 distinct species of trees in an area in which there were originally only four (Sawadogo *et al.* 2001, 41). Sawadogo's Zaï reinvented process stands as an example of effective solutions at the local level, based on traditional horticultural practices, without significant costs that contribute to diminishing poverty. It challenges the conventional idea of progress, since it is neither linear and ascendant, nor based on invention and innovation. It can have its ups and downs, and it is not based on the use and re-creation of already existing technologies.

Like Bouffier, Sawadogo changed the landscape and improved people's lives. He promoted biannual market days at his farm where he encouraged not only the exchange of seeds but also the exchange of knowledge. Sawadogo created a forest that acted as a springboard to motivate other farmers to plant trees, and, through his actions, he helped improve the lives of thousands of people in the Sahel region. This case study surpasses Giono's story of hope. It is real, not fiction.

Sawadogo is the leading example put forward by every environmentalist or ecologist for "greening the desert" (Kickul and Lyons 2016), and some even refer to him as the leader of a different kind of "green revolution" (Reij, Tappan and Smale 2009). Sawadogo's inspiring story lies behind the United Nations Convention to Combat Desertification's (UNCCD) decision to enhance the value of traditional knowledge and practices rather than modern technology to fight land degradation and desertification (Imeson 2012, 41).

Sawadogo's actions changed the climate at a local level. In the book *Drawdown*, published in 2018, Sawadogo appears as an example of climate change, since tree intercropping was proved to increase the region's rainfall as the ultimate consequence of the chain reaction caused by his sustainable horticultural technique, and inspired the climate scientist Jonathan Foley to highlight the "capacity of human beings to solve incredible challenges" (Foley in Hawken and Steyer 2018).

In the end, Sawadogo's actions were undermined by political and economic factors. Following release of the documentary, Sawadogo gained international reputation; he visited more than 20 countries to talk about his experiences, and, in 2010, was invited to speak at the International Conference in South Korea to Combat Desertification, organized by UNCCD. Unfortunately, and after so much success, Burkina Faso's government seized part of his forest for other purposes, offering him minimal rewards for land he had inherited from his father and to which he had added so much value.

This singular case shows that there is at least knowledge to combat desertification at a local level. Traditional horticultural practices reinvented through a technology-in-use framework proved to be the most effective, to such an extent that climate change occurred. Nevertheless, its success did not surpass the local scale due to political and economic factors. Moreover, when extensive reforestation efforts to fight desertification have been implemented at national and continental scales such as in the Great Plains, USA, the Gobi Desert, China or the Sahara, Africa, political and economic factors constrain their outcomes in equal measures.

The Great Plains Shelterbelt

The Great Plains Shelterbelt was the first large-scale planting of tree windbreaks from North Dakota to northern Texas, covering an area 1,000 miles long and 100 miles wide. It occurred after the Dust Bowl in the 1930s, and resulted in significant soil erosion, drought and the devastation of entire villages by sand.

This region was one of the windiest in the USA, and severe 60-day-droughts were common, to such an extent that agriculture was not a productive economic activity and vegetation was limited to native plants that do not need a regular water supply. It had been like this until the beginning of the twentieth century. However, between the two World Wars, the soaring prices of cereals stimulated farmers to plant more cereal fields in this region, despite the harsh conditions. Consequently, deforestation occurred to such an extent that between 1910 and 1920 the growth of cereals duplicated in the southern plains (Geist 2017). Therefore, the consequences of the severe droughts and dust bowls that occurred in the 1930s were worse than previously. When farmers saw their fields destroyed they abandoned the region, and the situation became a national issue. In Geist's opinion, the Great Plains Shelterbelt would never have occurred had farmers from former decades tried to transform their lands into productive areas (Geist 2017).

President Franklin Roosevelt (1882–1945) himself addressed the problem. He was the main promoter of the Great Plains Shelterbelt and the creation of the Soil Erosion Service. Since it was the first time that a mega-curtain was to be cultivated, the project was itself a forest, climate and environmental laboratory. Experimentation took place, specifically regarding the choice of tree species and of plantation sites. Decisions on the implementation of the project occurred as fast as the research took place.

The plan of the Shelterbelt consisted of 100 bands of woodland, separated by a mile of farmland to be cultivated by men in need of relief, and supervised by the United States Forest Service. The choice of trees privileged small and slow-growing species, such as the green ash, hackberry, Chinese elm, honey locust and osage. The location of the Shelterbelt followed the incidence of the winds: the stronger they were, the larger the Shelterbelt had to be. Still, due to its first critics the Shelterbelt shifted eastwards, to the area of tall grass where the chances of trees growing were considered to be much higher. The proposed Shelterbelt was also criticized for being inefficient for farming protection. Experts argued that only if farmland was one-tenth of a mile, rather than a mile, could the tree-planting effectively protect the agricultural fields. However, with this dimension it would be difficult to use farm machinery (Visher 1935, 68).

Still, most debate, then as well as today, was about the impact on climate change. President Franklin Roosevelt and his team advocated that it would alter the climate by changing soil and atmospheric conditions, increasing air humidity, and, consequently, rainfall. Moreover, planting trees would also create a barrier against winds (Visher 1935, 63). At the same time as the project progressed at a slower pace, scientific research was carried out and several qualified experts arranged conferences and published their own considerations on the feasibility of the project.[6] None of them believed that the Shelterbelt could modify the climate to such an extent and they were also quite sceptical that the project would be successfully delivered, doubting that millions of plants could be cultivated and grown (Visher 1935, 64). In 1935, Stephen Visher discussed the climatic changes claimed to be brought about by the Shelterbelt (Visher

1935), based on the theoretical framework put forward in *Climate Changes, Their Nature and Causes* (1924). In 1935, Visher concluded that "it is by no means clear that the shelter belt would yield an increased rainfall" (Visher 1935, 73). Moreover, he states that the Shelterbelt "would not produce the improvement of climate claimed for it by the advocates, and that climatically it is not justified" (Visher 1935, 73).

As more than 80 years have elapsed since then, we are in a better position to evaluate the project's impact on the climate. In 2007, Norman Rosenberg considered that there is no evidence to support the vision of President Franklin Roosevelt's Committee on the Future of the Great Plains in 1936. The climate did not change and rainfall did not increase, except at a local level (Rosenberg 2007, 24). However, the partial failure of this national project does not entitle one to deduce that growing a shelterbelt will not impact upon climate change. Many things went wrong. For example, the 1953 Report by Ernest George concluded that if methods for selecting, planting and caring for trees had been followed, this would have allowed for "reasonable assurance on most farms in the northern Great Plains of effective shelterbelt protection for many years" (George 1953, 55). Therefore, we can conclude that the choice of species and horticultural techniques did not follow the USA Forest Service's advice. Moreover, the project was jeopardized by tight budgets and extended drought, as well as by the deaths of many farmers who believed in the benefits of those shelterbelts.

Summing up, up until the twenty-first century, the Great Plains Shelterbelt was considered to be the most focused effort to address an environmental problem in the USA, although its outcomes did not meet their expectations and new dust storms continue to devastate the region (Wertz 2013). Irrespective of the scale, only some benefits were acknowledged at a local level. Moreover, it also contributed to solving a social problem in the 1930s. Employment was sparse, and this programme was responsible for hiring thousands of men, to such an extent that it became known as the "relief job". Notwithstanding the fact that the goal to change the climate was not achieved, we do not know how much of this was undermined by deficient implementation and political and economic factors.

The Green Great Wall of China

Governments have continued to envision gardening on a large scale through tree-planting as the solution to fight desertification and dust bowls. China has also tried it.

Only 2 per cent of China's original forests still survive today and over a quarter of its territory is currently covered with sand. Over past decades, reforestation has been considered as a major task by China's governments. Moreover, a range of challenges have been raised by the Gobi and Kubuqi deserts in China, since the dust from sandstorms, known as the "yellow dragon", has been damaging major cities such as Beijing every spring (Heshmati and Squires 2013, 3–74).

To face this problem a systematic programme of tree cultivation is taking place in northwest China along the Gobi Desert and the Ming portion of the Great Wall, from the western region of Xinjiang to Heilongjiang. It aims to limit the expansion of the desert and protect the cities under its influence. Inspired by the Great Wall, this major project to fight the spread of the Gobi Desert was named the "Green Great Wall of China", although, when completed, it will not be envisioned from the Moon as a wall but rather as a huge green spot, similar to the one in Amazonia.

Work began in 1978 and is projected to last until 2050. The plan is to cultivate 26 billion trees and grasses to cover an area 2,800 miles (4,500 km) long. Since the 1970s forest cover has already increased in China, from 12 to 16 per cent. The objective is to cover 42 per cent of the territory by 2050 and to increase the planet's forest by more than one-tenth. To achieve the goals established for 2050, two trees per citizen per year have to be cultivated and the whole population contributes to this effort through high school activities for all children over the age of 11.[7] In China the Green Great Wall is being built by forced unpaid volunteers, contrary to the former case where the US government paid farmers, contributing in this way to solving a between-wars socio-economic problem.

Since it is still ongoing, this project is difficult to grasp, and opinions are divided. Some are quite sceptical that it will influence climate change, while others, including its promoters, argue that changes can already be perceived.

Even if at a national level China is in fact able to stop desertification, this project will not compensate for planetary carbon emissions at a global scale,[8] and China remains the biggest timber consumer responsible for the inflow of illegal timber, while at the same time promoting deforestation in other parts of the world.[9]

In 2015, the *Telegraph* announced that the poor quality of China's Green Great Wall would mean that it would have a limited impact on climate change (Connor 2015). Even though the colossal scheme of tree-planting is fast growing, "they are poor quality and are regularly destroyed by storms. Scientists also claim they do little to nurture the growth of biodiversity" (Connor 2015). Nevertheless, the benefits of this wall are already felt in the terrain as Beijing is being spared devastating dust bowls, and the Great Green Wall of China is already envisioned as the ecologic counterpart of China's Great Wall.

Drawing upon this Green Great Wall, a continental project is underway, and the Great Green Wall of Africa is already expected to stop the spread of the Sahara Desert.

The most important horticultural project of the planet: the Great Green Wall of Africa

Between the Sahara Desert and the savannah, featuring grassland plains, lies a fringe of arid land known as the Sahel. The word "Sahel" derives from the Arabic *sahil* which means "shore", evoking the vision of a green land which many travellers had when coming from the desert. The Sahel was a line of scrub, bush, thorns, sandy soils and small, sporadic trees. However, severe droughts in the 1970s transformed it into a desert-like region. "Desertification will be associated by many people with the Sahel region and with the droughts that occurred in the 1970s when millions of people perished. At that time climate was considered a major factor in desertification" (Imeson 2012, 36). However, the aftermath of the great Sahelian drought, associated with famine and the deaths of millions of people, not only proved that desertification "is triggered by anthropogenic disturbances" (Puigdefábregas 1995, 311) but it also became a global issue in the prospect of a global food crisis. Although it took 15 years to be created, the United Nations Conference on Desertification (UNCOD) is an outcome of the dynamics generated in the aftermath of the Sahelian drought.

The Great Green Wall of Africa became the most important horticultural project of the planet to stop the current desertification trend. If it fails, two-thirds of arable land could be lost. Hence, 20 countries of the sub-Saharan region, headed by Senegal, are united to reverse desertification in a region between the Sahara Desert and tropical forests, which were at one time fertile and suitable for the growth of crops. The idea was formulated by the British scientist Richard St Barbe Baker in the 1950s and brought back to life by the President of Nigeria, Olusegun Obasanjo, in 2005. The Great Green Wall of Africa was endorsed by the Assembly of the African Union and comprises not only the planting of a transcontinental shelterbelt, but also supports local projects on agroforestry.

The Great Green Wall of Africa stands as one of the most promising resolutions on the planet's afforestation. It comprises an area between Dakar and Djibouti 15 km wide and 7,500 km long. It is hoped that this wall of trees will change the soil – it aims to attack its degradation – and humidity in the air, inverting the desertification process begun with the Sahelian drought in 1968 by changing the natural biophysical conditions. The choice of tree varieties was very important and the United Kingdom has participated with its Kew Gardens expertise.

In 2016, at the International Conference on Climate Change held in Paris, this "grand vision for a wall of vegetation reaching thousands of miles across Africa and slowing the spread of the Sahara Desert took a step closer to becoming a reality" (Laing 2016). The project, which was adopted by the African Union in 2005, was boosted in December by $4 billion of funding from signatories to the historic UN Climate Change Summit in Paris (COP21) and gathered more funding from the World Bank and the French government, one of the continent's major former colonizers (Laing 2016).

Elvis Paul Tangam, the African Union Commissioner in charge of the plan, stated that more than 27,000 hectares of indigenous trees, amounting to 15 per cent of the total plantation, had already been planted, largely in Senegal and Burkina Faso, with significant impact on environmental and ecological biodiversity, as many animals, such as antelopes, hares and birds, which had disappeared from those regions for 50 years, had reappeared (Laing 2016).

Studies have shown how land degradation in Europe is associated with the soil and ecosystem's degradation, erosion, salinization, as well as wildfires, climate change, land-use practices, pollution, contamination and compaction (Do Ó and Roxo 2009). Pasternak and Schlissel take as a starting point the Mediterranean region and the importance of the olive tree, as one of the oldest and most successful rain-fed systems in the world, to stress the importance of the choice of plants to fight desertification (Pasternak and Schlissel 2011, 20). In Portugal, studies have demonstrated how the transformation of Montado, Alentejo into wheat fields in an unsustainable way has contributed to land degradation (Roxo and Mourão 1994). Recent studies have also demonstrated that the lack of sustainability of the Algarvean landscape is related to its lack of identity, and that the choice of autochthonous plants could promote a wise use of water and the environment in both its cultural and ecological facets (Rodrigues 2017). The choice of species for the Great Green Wall of Africa shares this rationale.

Pasternak and Schlissel propose systematic tree-planting for the African dry lands of *Zizyplus Mauritania*, *Vitellaria paradoxa* and *Balanites aegyptiaca*. The first two are native to semi-arid regions in Africa and *Balanites aegyptiaca* is a colonizing plant, but can act as an agent for rehabilitation, since it is among the first species to grow on degraded soils (Pasternak and Schlissel 2011, 25–26). For the time being, many indigenous trees have already been planted to maximize shade and prevent groundwater loss, and shrubs that can be grazed by livestock, including acacia trees, from which they can gain some extra economic profit. For example, the Tunisian Sarah Toumi promoted the growth of acacias to form grids among the usual crops to such an extent that she became known as "La Dame aux Acacias" (Henry and Tubiana 2018). Some of the agroforestry projects developed in the region employ women in order to empower both them and the local population.

This case study is not merely about how the success of afforestation could benefit humankind, but also about the active role of vital resources such as water and soil, since they either foster or hinder the growth of trees. Man–forest interactions have been treated instrumentally – viewed in terms of how certain groups of people became hegemonic through the use and control of forest resources. Much less studied are processes whereby forests and trees as subjects come to exercise a certain kind of agency in social contexts, for example, in forcing the decision to grow or not to grow. It is not enough to take into consideration human action. The dry lands of Africa cannot rely solely on rainfall due to its unreliability, and action should consider groundwater supply, as it is estimated that deep aquifers lie under the semi-arid and arid countries of western Africa (Pasternak and Schlissel 2011, 21). This sort of non-human agency matters, it has practical consequences, and it must therefore be properly taken into account if we are to advance towards sustainable socio-political-environmental projects.

Conclusion

This chapter discusses various case studies which help rethink solutions based on innovation, both global and Western/European centred, by emphasizing traditional horticultural practices, recourse to old knowledge and technology-in-use, and rethinking the relations between the local and the global. In every case, expertise in the selection of species, traditional horticultural techniques and irrigation methods lies behind the solutions.

Although these case studies stand for the good Anthropocene, there is more evidence of the results at a local than at a global scale. These projects and achievements have helped solve the problems of the "garden", not of the "planetary garden". The *homo-gardinus* attitude sought to avoid desertification of the Sahel region and the expansion the Gobi and Sahara deserts. Despite these victories at a local level, no matter whether regional, national or continental, they did not have a positive global impact, as climate changes continue to worsen.

However, on the contrary, the local might be influenced by the global, as every negative effect on climate change and ecosystems on one side of the world can affect the other side. Fortunately, resilience and resistance avoid resources going from the local to the global, and that is what "allows the functions and ecosystems to be provided" (Imeson 2012, 27). But even at a local scale, environmental results are being challenged by corrupt political powers. The UNCCD has recently reviewed its strategy and proposed the next decade as one dedicated to desertification. Moreover, it evaluates success through how the lives of people have improved in the short term rather than only through the reduction of land degradation (Imeson 2012, 10).

However, and even if results at a local scale constitute positive evidence, we cannot give up on the ambition to achieve success at a global scale. The problem is global and must have a solution at this scale. Drawing upon Miller Hillis, Colebrook and Cohen's *Twilight of the Anthropocene Idols* who stated that we should "stage the revolution, return to profound thinking, reinvent the subject, and recognize ourselves fully as one global humanity", I argue that the solution to this anthropogenic crisis can only come from cross-boundary solutions between the humanities and the sciences. The history of gardens and landscapes can be especially powerful tools for rethinking science, technology and the environment.

The epistemological framework of gardening may be the key to this problem, and an attitude of the *homo-gardinus* the right one to take care of a garden as big as the planet. Nevertheless, in the *homo-economicus* world this is neither obvious nor easy. Market rules will certainly not restore balance and equity between the local and the global, between Third World Countries and Developed Countries. The challenge for the humanities is to understand the environmental perspective as fundamental, far beyond market rules and political interests, by contributing to a vision of the planet *qua* garden, and of humankind *qua* gardener, whose actions are directly intertwined, benefiting or biasing both.

Acknowledgement

I thank the referees for their comments and suggestions. FCT MCTES Project IF/00322/2014 and PEst-OE/HIS/UI0286/2014.

Notes

1 Mission Life Force foresees the inclusion of Ecocide as the fifth International Crime Against Peace as follows:

> Ecocide is the extensive damage to, destruction of or loss of ecosystem(s) of a given territory, whether by human agency or by other causes, to such an extent that peaceful enjoyment by the inhabitants of that territory has been or will be severely diminished.
> (Consulted online on the webpage "Eradicating Ecocide", 27 February 2018)

The proposal was presented to amend the Rome Statute in 2010. See http://eradicating ecocide.com/the-law/what-is-ecocide.

2 For example, such was the case of Mediterranean civilizations. Adopting a certain nostalgia for those times when culture and nature lived in harmony, James H.S. McGregor pleas for "Back to the garden" (McGregor 2015).

3 Among the large bibliography on the *homo-economicus*, I point out the work by Merve Gulacan, as he develops the concept of the *homo-economicus* in which individual behaviour is controlled by rationality and is driven for the maximization of economic gain (Gulacan 2015, 2).

4 "Nature" derives from the Latin verb *nasci* (to be born), homologous to the Greek *physein* (to be generated); thus, the etymology of the word prompts an intimate connection between totality and essentiality.

5 The trailer may be seen at www.1080films.co.uk/Yacoubamovie/. This film won two international prizes in 2010.

6 These articles were: Butler 1937; and the talk by Ellsworth Huntington, Yale University, at a meeting of the Society for the Protection of New Hampshire Forests, at Plymouth, in 1934 (Droze 1977, 86).

7 "'Great Green Wall' and 'Green Corps': Joining with Our Chinese Youth to Stop the Spread of Deserts". *International Understanding*, 2014, Issue 1, pp. 46–49.

8 As the Environmental Investigation Agency's (EIA) member Faith Doherty considers (Connor 2015).

9 China does not legislate against the inflow of illegal timber, "Unlike the EU Timber Regulation and the US Lacey Act – and relies solely on self-regulation within its timber industry" (Connor 2015).

Bibliography

Butler, Ovid. 1937. "Conservation at the Forks: Proposed Reorganization of the Federal Government Raises Questions Vital to Agriculture, Forestry and Other Fields of Conservation". *American Forests*, 43(3): 109.

Clément, Gilles. 2015. *"The Planetary Garden" and Other Writings*. Pennsylvania, PA: University of Pennsylvania Press.

Connor, Neil. 2015. "How China's 'Great Green Wall' Is Poor Quality and Will Have a Limited Impact on Climate Change". *Telegraph*, 29 November (consulted online 27 February 2018).

Do Ó, Afonso and Roxo, Maria José. 2009. "Drought Response and Mitigation in Mediterranean Irrigation Agriculture". *Water Resources Management Transactions on Ecology and the Environment*, 125: 515–524.

Droze, Wilmon Henry. 1977. *Trees, Prairies, and People: A History of Tree Planting in the Plains States*. Denton: Texas Woman's University.

Edgerton, David. 2006. *The Shock of the Old. Technology and Global History since 1900*. London: Profile Books.

Geist, Helmut. 2017. *The Causes and Progression of Desertification*. Abingdon, Oxon: Routledge.

George, Ernest. 1953. *Thirty-one-year Results in Growing Shelterbelts on the Northern Great Plains*. Washington, DC: U.S. Department of Agriculture.

Giono, Jean. 1953. *The Man Who Planted Trees* (online). www.idph.net (accessed 14 February 2018).

"Great Green Wall' and 'Green Corps': Joining with Our Chinese Youth to Stop the Spread of Deserts". 2014. *International Understanding*, 1: 46–49.

Gulacan, Merve. 2015. *The Concept of "Homo Economicus" and Experimental Game: Is "Homo Economicus" Still Alive?* Verlag: Grin.

Hanley, Paul. 2014. *Eleven*. Canada: Friesen Press.

Hawken, Paul and Steyer, Tom, eds. 2018. *Drawdown: The Most Comprehensive Plan Ever Proposed to Reverse Global Warming*. New York: Penguin Books.

Henry, Claude and Tubiana, Laurence. 2018. *Earth at Risk: Natural Capital and the Quest for Sustainability*. New York: Columbia University Press.

Heshmeti, G. Ali and Squires, Victor. 2013. *Combating Desertification in Asia, Africa and the Middle East: Proven Practices*. Dordrecht, Heidelberg, New York, London: Springer.

Imeson, Anton. 2012. *Desertification, Land Degradation and Sustainability*. Oxford and Hoboken: Wiley-Blackwell.

Kaboré, Daniel and Reij, Chris. 2004. *The Emergence and Spreading of an Improved Traditional Soil and Water Conservation Practice in Burkina Faso*. EPTD Discussion Paper No. 114. Washington, DC: Environment and Production Technology Division, International Food Policy Research Institute.

Kickul, Jill and Lyons, Thomas S. 2016. *Understanding Social Entrepreneurship: The Relentless Pursuit of Mission in an Ever Changing World*. New York: Routledge.

Laing, Aislinn. 2016. "Great Green Wall Thousands of Miles Long oCuld Be Built across Africa to Stop the Spread of the Sahara". *Telegraph*, 3 May.

Kotke, William H. 2005. *Garden Planet: The Present Phase Change of the Human Species*. Bloomington, MD: Authorhouse.

Latour, Bruno. 1986. *Science in Action. How to Follow Scientists and Engineers through Society*. Cambridge, MA: Harvard University Press.

Latour, Bruno. 2004. *Politics of Nature. How to Bring the Sciences into Democracy*. Cambridge, MA, and London: Harvard University Press.

Lawrence, Derek. 1972. "The Ideological Writings of Jean Giono (1937–1946)". *The French Review*, 45(3): 588–595.

McGregor, James Harvey. 2015. *Back to the Garden: Nature and the Mediterranean World from Prehistory to the Present*. New Haven, CT: Yale University Press.

Messervy, Julie Moir. 2007. *The Inward Garden. Creating a Place of Beauty and Meaning*. New York: Bunker Hill Publishing.

Ng, Irene C.L. and Lu-Ming Tseng. 2008. "Learning to be Sociable – The Evolution of Homo Economicus". *The American Journal of Economics and Sociology*, 67(2): 265–286.

Pasternak, D. and Schlissel, A. 2011. *Combating Desertification with Plants*. New York: Kluwer.

Ponting, Clive. 2011. *A Green History of the World: The Environment and the Collapse of Great Civilizations*. London: Vintage Books.

Puigdefábregas, Juan. 1995. "Desertification: Stress beyond Resilience, Exploring a Unifying Process Structure". *Ambio*, 24(5): 311–313.

Reij, Chris, Tappan, Gray and Smale, Melinda. 2009. *Agroenvironmental Transformation in the Sahel: Another Kind of "Green Revolution"*. International Food Policy Research Institute, 2020 Vision Initiative.

Rodrigues, Ana Duarte. 2017. "Sustainable Beauty for Algarvean Gardens: Cross-boundary Solutions between the Humanities and the Sciences". *Interdisciplinary Science Reviews*, 42(3): 296–308.

Rosenberg, Norman J. 2007. *A Biomass Future for the North American Great Plains: Toward Sustainable Land Use and Mitigation of Greenhouse Warming*. The Netherlands: Springer.

Roxo, Maria José and Mourão, Jorge Manuel. 1994. *Land Degradation in the South Interior Alentejo-Mértola Region: Historical Overview of Agricultural Impacts on the Environment*. Lisbon: New University of Lisbon.

Sawadogo, Hamado, Hien, Fidèle, Sohoro, Adama and Kambou, Frédéric. 2001. "Pits for Trees: How Farmers in Semi-arid Burkina Faso Increase and Diversify Plant Biomass". In *Farmer Innovation in Africa: A Source of Inspiration for Agricultural Development*, edited by Chris Reij and Ann Waters-Bayer (pp. 35–46). London: Earthscan.

Scruton, Roger. 2014. *How to Think Seriously about the Planet: The Case for an Environmental Conservatism*. Oxford: Oxford University Press.

Sing C. Chew. 2001. *World Ecological Degradation: Accumulation, Urbanization, and Deforestation 3000 BC–AD 2000*. Walnut Creek, Lanham, New York, Oxford: Altamira Press.

Visher, Stephen S. 1935. "Climatic Effects of the Proposed Wooded Shelter Belt in the Great Plains". *Annals of the Association of American Geographers*, 25 (2): 63–73.

Wertz, Joe. 2013. "Dust Bowl Worries Swirl Up As Shelterbelt Buckles". *NPR*, 10 September. www.npr.org/2013/09/10/220725737/dust-bowl-worries-swirl-up-as-shelterbelt-buckles.

Zylinska, Joanna. 2014. *Minimal Ethics for the Anthropocene (Critical Climate Change)*. Michigan: Michigan Publishing.

6 From *Homo faber* to *Homo hortensis*

Gardening techniques in the Anthropocene

Astrid Schwarz

Gardens and garden practices are identified with material and social techniques that are loaded with values like care and responsibility, with beauty and benefit, and of course with questions of lifestyle, of comfort and design. All of these values seem to lend themselves to reflect upon practices of maintenance and sustainability and to develop "options of adaption and mitigation" for future societies, altogether commonplace in the climate change debate. For about the past 15 years, the label "Anthropocene" has been used to highlight the urgency of managing planetary boundaries (Crutzen 2002, Rockström *et al.* 2009). Increasingly, the concept was identified with concern about the occurring changes in the human condition and debated in an interdisciplinary context (Sloterdijk *et al.* 2011, Palsson *et al.* 2013, Smith *et al.* 2013, Oxman 2016). However, thus far the negotiations on climate change have been dominated to a great extent by political deliberations about an adequate interpretation of complex scientific models (Schneider *et al.* 2002) on the one hand, and on the other by searching for technoscientific solutions on a large scale, including the earth system as a whole. Accordingly, geoengineers have proposed deploying nanoparticles in the stratosphere to reduce solar radiation (Keith 2013) or to fertilize the oceans to increase carbon turnover of algae and finally to use the deep sea as a carbon sink (Smetacek *et al.* 2012). Interestingly enough, the proponents of this merely technoscientific approach describe it in terms of a "gardening business" (e.g. Keith 2009); what is more, in general the discourse seems to move from the language game of a stewardship of planet Earth to a language game of garden management. Inversely, the traditional gardening community starts to ask questions about the influence of literal gardening practices on climate change (e.g. Cook 2008). In 2015, the Royal Horticultural Society launched a five-year plan "to meet the challenge of climate change and help gardeners be more eco-friendly" (RHS 2015); that is, to better understand the critical role of gardening at large, the management of plants, and its impact on the environment for the benefit of people and the planet.

The vantage point of a *Homo hortensis* is expected to help, first, to shift attention to historically available gardening techniques and concepts through the history and theory of gardening (Hunt 2000, Henderson 2004, Turner 2004), and second, to focus more on considerations about epistemic and normative capacities at hand being already implemented and debated in different language games of gardening.

Accordingly, the general thesis of this chapter is that *Homo hortensis* disposes of forms of life that provide gardening techniques and contribute to work on local problems in practice and at the same time afford a sociotechnical imaginery of global conditions in the Anthropocene.

Who are *Homo faber* and *Homo hortensis?*

Homo hortensis is to some extent conceived of as a refinement of *Homo faber*, the skilled craftsman of our technosphere, the technological systems of the Anthropocene (Haff 2014). *Homo faber* is a widespread figure in the literature of the history and philosophy of science and technology as well as in the fine arts representing the modern man that creates tools to act in the world. *Homo faber* stands for the tremendous success of the happy liaison between science and technology. Hannah Arendt has characterized her *Homo faber* by figuring out its particular mode of acting in the world and by illuminating the closeness of experimenting and of producing. She pointed out that experimental activities are in themselves a mode of fabricating. "The experiment repeats the natural process, as if it were a matter of concern to produce the things of nature once again" (Arendt 1981, 288). In the history of the natural sciences this *Homo faber* marks an important shift from "why" questions to "how" questions. The modern conjunction of producing and understanding was famously characterized by the Kantian dictum "give me matter and I will build a world out of it" (Kant 1996), recently updated with the phrase "give me a laboratory and I will raise the world" (Latour 1983). Productivity and creative ingenuity are therefore the idols of *Homo faber* and its "fabricating making" as Arendt puts it. In her conceptual system, *Homo faber* is only distinguished from *Animal laborans*, the labouring body, in that it has a more definite practical intelligence and better manufacturing skills. On the other hand, *Homo faber* is also clearly different from *Zoon politikon*. This mode of acting offers, according to Arendt, the most fruitful way for human beings to realize their freedom in an active existence by taking political action. The mode of existence of *Homo hortensis* and the characterization of its relationship to the furniture of the word builds on the three modes of acting, discussed above.

Homo hortensis means the gardening man, or, in a more straightforward translation, the one who belongs to the garden. Accordingly, the form of life of *Homo hortensis* is not just working or being in the garden, but organizing its life following the principles of gardening. *Homo hortensis* cultivates its environment in terms of a garden, while he or she lives in urban or rural communities inspired by gardening ideals and uses garden technologies. This also means that gardens never exist independently of the world shaped by human action – even if they are not fully embraced by or contained in this world.

Table 6.1 The fourth mode of action, assembling/designing, draws upon the three modes of action developed by Hanna Arendt in her book *Vita Activa* (1981)

Mode of action	working/ labouring	producing/ fabricating	reasoning/ deliberating	assembling/ designing
Character	*Ponos* *animal laborans*	*Poiesis* *Homo faber*	*Praxis* *Zoon politicon*	*Homo hortensis* with nature
Imaginery of nature	necessities in nature	disposition of nature	beyond nature	manipulative/ cooperative
Temporal order	cyclical (indefinitely)	linear/directed (temporary)	process-related (both)	process-related (both)
Spatial order	housekeeping *Oikos* (home, factory)	shop, laboratory	*Agora*, public	public sphere, field laboratory
Products	consumption objects	objects of utility	codes of conduct, regulation	garden work
A garden is	not necessary for survival	an object in transition	socially/ materially construed	cooperative/ co-productive

For instance, the design of gardens can be shaped by political intentions – think of the French baroque gardens as a representation of absolutism (Mukerji 2002), of the English landscape garden referring to the ideals of English parliamentarianism (Turner 2004), or of contemporary social gardens materializing ideas about participation in decision-making, or the care of future generations (Schwarz 2016). In his book *Gardens: An Essay on the Human Condition*, Robert Harrison emphasizes that "all nonimaginary gardens have in common that they come into being through human agency. The fact that this is self-evident does not make it any less decisive" (Harrison 2008, 65).

Garden practices in the Anthropocene

In the face of the radical novelty of the Anthropocene, the human being as gardener is not new and has been described as such again and again. Gardening may be regarded as a human condition; it is linked to famous concepts such as the Garden of Eden, or utopia, but also, perhaps more astonishing, to English parliamentarianism (the political landscape garden), or to experimentation. Thomas Morus presents us with a garden concept that might well pass as a forerunner of the contemporary community garden (Burrell and Dale 2002), while Francis Bacon advocates for the garden as an outdoor laboratory to afford more and bigger fruits, more efficient drugs, and a better life for all (Bacon 1627). All of these suggestions present the garden as a model to probe socio-economic configurations and of possible interactions with natural surroundings and resources.

However, the laboratory perspective also offers itself to be used for analytical purposes. A specific garden then becomes a controlled environment where artificial specimens yield generalities that are true anywhere, requiring of the analyst distance and objectivity. The garden may be regarded as a "truth-spot". This is a paraphrase of "City as Truth-Spot" of sociologist Thomas Gieryn (2006), discussing urban research in Chicago as a particular case to draw general conclusions about urban science.

The truth-spot "garden" may reveal technologies, epistemic values and an eco-political agency that may be used to better understand the role of human beings in the Anthropocene, particularly under the conditions of global warming; that is, under the conditions of a more or less radical change of their *oikos*, planet Earth.

In this sense, I assume that the garden provides a model to act on different scales, be it in an urban backyard, on the landscape level, or to develop an international policy for mitigation and adaptation to climate change. Accordingly, the opulent culture of gardening offers a rich toolbox to cultivate our environment in a sustainable way. Taking all of this into consideration, I want to argue that gardens afford a truth-spot to manage pieces of nature ("Naturstücke", after Böhme 1985) on different scales.

"Garden earth" and other sociotechnical imaginaries

The literature in the natural and engineering sciences provides some evidence that the garden is seen as a challenging model. However, thus far it has been largely used in a metaphorical sense without spelling out the conceptual benefits. The garden works rather as an umbrella concept, embracing different, even controversial claims and assumptions and therefore also quite different ideas about the scientific and technological practices that the model "garden" might afford. Furthermore, it is not only the language game of gardening that is used to characterize the feelings of unease which accompany the changing conditions for humankind and its dwelling place, planet Earth, in the age of the Anthropocene. There are different pictorial and narrative visions around focusing on at least four different sociotechnical imaginaries (Jasanoff and Kim 2015): spaceship earth, the earth as self-regulating Gaia, patient earth, and finally garden earth.

Spaceship earth became popular in the 1960s during the era of the space race. One of its promoters was architect Buckminster Fuller, known for his vision of comprehensive planetary planning (Fuller 1968, Höhler and Luks 2006). His stance was that technology and know-how already exist, so that humanity can successfully surmount global challenges. Until today, spaceship earth has accumulated rich meaning.

The Gaia hypothesis as a sociotechnical imaginary is also an outcome of the 1960s; it states that the atmosphere and surface sediments of planet Earth form a self-regulating physiological system. In the strong version of the hypothesis, the Earth itself is a self-regulating organism (Lovelock 1979). This was rejected by the scientific community; however, weaker forms of the hypothesis are still under discussion.

Patient earth is a popular, more recent sociotechnical imaginary inspired by medical culture. It is dominated by ideas about how a self-organized unit works, how the concepts of health and disease are related, and how the patient is to be monitored most efficiently (e.g. Plag 2016). In science policy, the health issue is an almost irresistible imaginery; it fuels an economy of promises that in a strikingly effortless manner encompasses the mesocosmic world of the human body.

Garden earth as a sociotechnical imaginary is different in that it neither comes along with an organism analogy, nor does it refer to a merely technical artefact. Instead, garden earth refers to a cooperative work or management strategies that are always characterized by concepts of interrelatedness. Some language games involve a complementarity and reciprocity of natural and cultural structures and processes, of a gardener susceptible to the garden's order (Ihde 1990, Cameron 2014). Others refer to the idea of the garden as a system to be controlled and classified, in which the gardener resembles the Promethean engineer (Keith 2009). Thus, the garden imaginary is remarkably diverse; what is more it tells a thick story in terms of historical background and grounded practices all over the world.

Similarities and dissimilarities of the sociotechnical imagineries

What ties all these imagineries together is that they refer to big ideas being rooted in the humanist tradition such as human rights, a fair distribution of wealth, good health and well-being for all, the sustainable use of natural resources, sustainable cities and communities, etc., all of these being values also listed, for instance, in the sustainable development goals by UNESCO (https://en.unesco.org/sdgs). It is almost impossible not to agree without being denounced as a kind of "immoral" person. Accordingly, the values and virtues involved come with an almost irresistible lure of the yes and the appeal to all of us to join the camp of do-gooders. Over the years, these universalities and their international institutional settings have been challenged by STS scholars, political scientists and philosophers in the sense that it is not enough to rely on climate change deliberations on the level of international policy or a nation's efforts to develop and settle regulatory frameworks. Instead, it was considered important to address the scale of human perturbation and its counteraction on a more local level, such as regions, cities and particularly the lifestyles and corresponding practices of citizens. In this context, the concept of "citizen science" has become an important keyword, as it was put forward, for instance, by a group of European political and environmental scientists, asking for governments "to provide an enabling and regulatory framework that supports the actions new agents of change are taking" in our societies (Hajer *et al.* 2015). Accordingly, citizen science reconceptualizes the concept of citizen by inserting it within a multiple role model. Thus, citizens are participants, critics and knowledge creators, and they perceive themselves as being engaged in knowledge production, a kind of collective reasoning (Felt and Wynne 2007) or user-driven innovation for sustainability.

Taking gardens and its gardeners seriously

In the following I argue that the sociotechnical imaginary of gardening is particularly apt in pursuing these different upstream strategies of the "new agents of change" in the Anthropocene. Taking the gardening imaginary seriously in this sense may support a reflexive framework for *Homo hortensis*. What I expect from taking garden practices seriously and thus the conceptualization of *Homo hortensis* in contrast to *Homo faber* is the following.

First, gardens may be understood as epistemic objects that provide an experimental setting to learn about anthropocenic technics and to tackle them with different kinds of transformations – cultural, technical, environmental. This is why second gardens may be considered as skilful (kunstvoll) and artistic (gestalterisch) works affording a particular constellation of sociotechnical practices and a whole range of more or less natural artefacts, of a more or less cultivated nature. Third, it is the very activity of gardening that is able to induce rules, habits and in the longer run also lifestyles, thereby pursuing certain virtues and values such as modesty and humility, as a pre-condition of self-mastery, care and respect, and, closely related, even hope, as garden philosopher Gregory Cooper (2006) put forward by emphasizing that hope is not simply induced by this or that garden practice but pervades the very ethos of gardening. Ornamental gardens have been considered, just as the theatre in Schiller's writing, as being a moral institution. The design of a garden is considered as being a good deed and the constructed garden work affords morally good actions.

To make a garden is to engage in a planned, demanding, long-term enterprise (Cooper 2006, 96f.) and in this sense gardeners "live for the future" (Capek 2001, 167).

The garden is a protected space, the *Hortus conclusus* a distinct form, a cut-out piece of nature chosen for giving it human dimensions in an otherwise unbounded nature or urban jungle. This creates different relationships among the objects within this area, which also fosters a different perception of the gardener. The garden affords a distinct place of attention, of particular encounters, supported by values such as care, respect and contemplation.

Generally speaking, garden works may be regarded as a product of people's activities and imaginations, as a well-defined space that needs to be permanently maintained and simultaneously reinterpreted time and again – just as do works of technology.

In addition, gardens have always been perceived as the result of human activity. A garden is neither art nor nature; it is art-and-nature: "the garden is peculiarly dependent on the co-operation of nature" (Cooper 2006, I). Furthermore, there is an immense quantity of historical material available to learn about these cooperative nature–culture configurations. Since the eighteenth century, garden works have been explicitly included as a topic in philosophical discourse; however, discussions about garden works, how they should be assembled, what kinds of techniques are needed to design them, why they are valuable and what expectations and feelings they evoke started much earlier.

"Garden earth" as an inclusive language game

Particularly because gardening imaginary is used in such different language games it is of pertinent interest, because it allows one to follow and identify different strategies and ontological commitments; that is, epistemic–ontological relatedness. As has already been pointed out, garden earth is by no means exclusively used to herald the new agents of change, the application of green technologies and upstream policy. It is also suggested to introduce and manage major centralized technologies that rather may be identified with a promethean imaginary. Some geoengineers, for instance, suggest releasing nanoparticle clouds or installing huge mirrors in the stratosphere to mitigate climate warming, calling these practices a "gardening business" (Keith 2009, 2013). Others, namely more systems science-oriented authors, expect that the mode of gardening "provides an understanding of the complexity of the whole system" (Leinfelder 2013). In addition, the role of the gardener as steward is seen quite differently. For some authors, "garden earth" provides a role model of modesty in not claiming any ownership of garden earth (Cameron 2014); the planet is only leased and accordingly should be handled more respectfully. Other stewards are convinced of a need for planetary design and maintenance, putting some emphasis on a principle of care being implanted in the precautionary principle (Poel and Royakkers 2011, 102–104).

Obviously, in the garden imaginary there are conflicting attitudes and also practices at work that are accompanied by a different relatedness of ontological and epistemic interests and expectations.

The narrative thickness of garden works and their impact

To create or maintain garden works, very different activities are at stake, such as drafting, planning and constructing, hoeing, digging, seeding and planting, nursing and nurturing, manufacturing and gathering things together. Gardens are demanding in that they imply a temporal and spatial structure; they are in need of maintenance and regular services, otherwise they experience a change or even loss of their character. Garden works are the result of practices and techniques developed to deal with more or less cultivated and more or less artificial things; they offer a playful machinery of surprises and novelties. At the same time they can be identified with continuity and mainly the pleasurable recognition of order: in the eighteenth century for the representation of the laws of nature, in the eighteenth century for competing political models, and in the nineteenth century for the power of machines and industrial modernization. Gardens are distinct places where things may be connected in a systematic or completely unexpected way. These things are trees and bushes, waterlines and water basins, colourful flower beds and sheer green turf, sculptures, garden gnomes and ruins, walls and pathways. Of course, not all of these things are found in every garden, since gardens can be almost radically different in terms of form, features and intention, and it is difficult to define a common denominator. However, all garden works create relationships between particularities, and the garden may be considered as a model of so-called nested reciprocities. But studying a garden is a pertinent exercise to not only probe concepts of interrelatedness but to also become involved in the study of particular connections and one's own connectedness (Dumit 2014).

Garden works do not match that easily with a common model of a "work of technology"; on the other hand, they seem to be just as much an artwork than a work of technology, but also a work of art-and-nature. Garden works are characterized by their inter-position, an "inbetweeness" and ambivalence, and they are skilful works that embody scientific, technical and artistic knowledge.

In the following I will discuss a number of clusters of characteristics that emerge from garden works. I will tentatively probe certain modes of acting in the sense of gardening practices in the Anthropocene.

The first cluster is about the political dimension of garden works; they invite one to become involved and to engage in discussions about social conventions and underlying values. Garden practices raise, even induce, certain virtues; as garden philosopher Gregory Copper has pointed out, the garden is "hospitable" to various practices. Edible cities are perhaps a good example of this. Citizens engage in producing their own food and become sensitized to what they eat, where it comes from and how it is produced. In experiencing this collective gardening, they become involved in the deliberation of new food rules, for instance, about short transport distances, limitations of fertilizers, or the choice of locally adapted seeds, which can all translate into political awareness and action.

The second cluster is about resilience and robustness, and thus about strategies of how to tackle surprises and unexpected incidents. The garden is a place of change and the gardener must learn how to react in an adequate way to not completely lose the order and form of her or his garden. The garden may be perceived as a kind of local epistemic system permanently generating surprises that need to be handled by choosing adequate methods and technical means. Through permanent fitting and stretching, the objects of interest can be stabilized. In a book about garden techniques under the conditions of climate change, the author states, "alongside the problems, global warming does offer the gardener the opportunity to garden in an extended growing season and to experiment with a whole host of new and exciting plants" (Cook 2008, 14).

The third cluster is about the question of how the environment is perceived and which senses and technical means are reciprocally used. The answer to this question is crucial for any organism–environment system and its ongoing exchanges among the different levels of organization; reciprocity and attention are in a sense two sides of the same coin. The garden is a representation of order in that the temporalities of plants and animals structure the life of the gardener, who needs to cultivate them in the right spot at the right time. In this sense it is not human beings who lend an order to the garden, but it is the garden that defines the limits and capacities to transgress them. In his book *The Passionate Gardener* Rudolf Borchardt points out that order in the garden is the flipside of surveillance models in society insofar as they are effected in "education, and redemption, since these are the things with which all notions of order are concerned" (Borchardt 1951, 38).

A fourth cluster is about the particular perception of nature that garden work affords in that it calls "a pleasurable attention to human freedom", as Kant famously pointed out. Later, this argument was linked with the opportunity to enjoy contemplatively the extension of the terrain of landscape gardens, and again, some time later, an "Attention Restoration Model" (Kaplan 1995) was offered in psychological research claiming that there is a distinct psychological response to perceiving natural environments, such as scenic extent resonating with the need or desire of the moment. The so-called Biophilia hypothesis (Kellert and Wilson 1993) adopts a materialistic stance here; it says that people's desire for contact with nature has an underlying cause based on genetic fitness and competitive advantage: the natural environment is a vital resource for human well-being, both physical and mental.

The fifth and final cluster is grounded in numerous empirical studies showing that diversity in gardens, either in the city or in the countryside, is perceived as being a positive value. People prefer flower meadows and colourful borders to unvaried environments; in this sense diverse surroundings influence health and well-being in a measurable way. However, there is also a more theoretical argument here about the experience of otherness and of how to tackle the new, unknown, perhaps alarming challenges. Provided that the garden offers the experience of cooperation with nature, changing conditions in the garden may be expected to result in pleasurable activities, and not in a battle against an antagonistic and hostile nature. This can be traced easily in the garden literature, for instance, in a publication by the Royal Horticultural Society, London, in which two modalities of gardening are described. While the way of "old gardening" tends to shut out the world and changes the soil, erects barriers to reduce the wind and so on, the new gardening method focuses on the opportunities: "the challenge is to use the conditions on offer to best effect, with some minor adjustments" (RHS 2015).

Concluding remarks

The mode of gardening allows one to design and understand complex systems and at the same time involves actors in a cooperative way. *Homo hortensis* provides a different logic of linking epistemic and normative capacities: technology visioneering is not so much geared to the typical promethean imagineries of a futuristic drive and a belief in innovation forces but rather relies on works at hand having already been implemented, on use-based technologies (Edgerton

2007) and on an explicit reference to the historical archive of garden works, in which natural and cultural history have always been intertwined. Accordingly, garden works may be perceived as truth-spots to manage pieces of nature; likewise on the scale of landscape or a small backyard in the city. The two relational modes of generating artworks, composing and combining, that is, the composition of harmonized objects of an abstract order in contrast to the combination of surprising constellations by also using objects at hand, underline the different strategies of *Homo faber* and *Homo hortensis*.

Gardens provide an experimental setting to acquire local knowledge about ongoing environmental changes and to generate reasonable practices and suitable objects, also in terms of a citizen science. Gardening practices are embedded in a broader context of social and technical environments; they afford works on different scales, with different purposes and of radically different forms. Gardening is an activity that brings us into a close relationship with the objects and materials that surround us – plants, animals, water features, soil, sculptures, and so on. Gardening practices are profoundly cooperative in terms of overcoming the nature/culture divide. Gardening teaches us to be confident about the future, and a garden is cultivated with the future in mind. We engage in garden works hoping to be rewarded by the gifts of the garden; for instance, in a very mundane way by its fruits, but also by the growing and shaping of places of calm and beauty. Cultivating a terrain is a symmetrical relationship; by shaping and treating the terrain and its inhabitants, they induce rules and habits, and eventually influence our lifestyle.

Bibliography

Arendt, Hannah. 1981. *Vita Activa oder vom tätigen Leben*. Munich: Piper.

Bacon, Francis. 1627. New Atlantis. www.gutenberg.net.

Böhme, Gernot and Engelbert Schramm. 1985. *Soziale Naturwissenschaft. Wege zu einer Erweiterung der Ökologie*. Frankfurt am Main: Fischer Taschenbuch Verlag.

Borchardt, Rudolf. 1951. *Der leidenschaftliche Gärtner: Ein Gartenbuch*. Berlin: Matthes & Seitz.

Burrell, Gibson and Karen Dale. 2002. "Utopiary: Utopias, Gardens, and Organization". In *Utopia and Organization*, edited by Martin Parker (pp. 106–128). Oxford: Blackwell.

Cameron, W.S.K. 2014. "Conceiving the Earth Itself as Our Garden". In *Old World and New World Perspectives in Environmental Philosophy: Transatlantic Conversations*, edited by Martin Drenthen and Jozef Keulartz (pp. 53–71). Basel: Springer International Publishing.

Capek, Karel. 2001. *Das Jahr des Gärtners*. Berlin: Aufbau Verlag.

Cook, Ian 2008. *Waterwise Gardening*. London: New Holland Publishers.

Cooper, David E. 2006. *A Philosophy of Gardens*. Oxford: Clarendon Press.

Crutzen, Paul J. 2002. "Geology of Mankind". *Nature*, 415 (3 January): 23.

Dumit, Joseph. 2014. "Writing the Implosion: Teaching the World One Thing at a Time". *Cultural Anthropology*, 29(2): 344–362.

Felt, Ulrike and Brian Wynne. 2007. *European Commission: EUR 22700: Science and Governance: Taking European Knowledge Society Seriously*. Luxembourg: Office for Official Publications of the European Communities.

Fuller, Buckminster Richard. 1968. *Operating Manual for Spaceship Earth*. Carbondale: Southern Illinois University Press.

Gieryn, Thomas F. 2006. "City as Truth-spot: Laboratories and Field-sites in Urban Studies". *Social Studies of Science*, 36(1): 5–38.

Haff, Peter. 2014. "Humans and Technology in the Anthropocene: Six Rules". *The Anthropocene Review*, 1(2): 126–136.

Hajer, Maarten, Måns Nilsson, Kate Raworth, Peter Bakker, Frans Berkhout, Yvo de Boer, Johan Rockström, Kathrin Ludwig and Marcel Kok. 2015. "Beyond Cockpit-ism: Four Insights to Enhance the Transformative Potential of the Sustainable Development Goals". *Sustainability*, 7(2): 1651–1660. https://doi.org/10.3390/su7021651.

Harrison, Robert Pogue. 2008. *Gardens: An Essay on the Human Condition*. Chicago, IL: The University of Chicago Press.

Henderson, John. 2004. *The Roman Book of Gardening*. New York: Routledge.

Höhler, Sabine and Fred Luks, eds. 2006. *Beam Us Up, Boulding! 40 Jahre "Raumschiff Erde"*. Hamburg: Vereinigung für Ökologische Ökonomie.

Hunt, John Dixon. 2000. *Greater Perfection: The Practice of Garden Theory*. London: Thames & Hudson.

Ihde, John. 1990. *Technology and the Lifeworld. From Garden to Earth*. Bloomington, IN: Indiana University Press.

Jasanoff, Sheila, and Sang-Hyun Kim, eds. 2015. *Dreamscapes of Modernity: Sociotechnical Imaginaries and the Fabrication of Power*. Chicago, IL: The University of Chicago Press.

Kant, Immanuel. 1996. *Kritik der Urteilskraft*, edited by Wilhelm Weischedel. Frankfurt am Main: Suhrkamp.

Kaplan, Stephen. 1995. "The Restorative Benefits of Nature: Toward an Integrative Framework". *Journal of Environmental Psychology*, 16: 169–182.

Keith, David. 2009. "Is it Time to Consider Manipulating the Planet?" Interview by Jeff Goodell. *Yale Environment E360*, Yale School of Forestry & Environmental Studies, 7 January. https://e360.yale.edu/features/geoengineering_the_prospect_of_manipulating_the_planet.

Keith, David. 2013. *A Case for Climate Engineering*. Cambridge, MA: MIT Press.

Kellert, Stephen R. and Edward O. Wilson, eds. 1993. *Biophilia hypothesis*. Washington, DC: Island Press.

Latour, Bruno. 1983. "Give Me a Laboratory and I Will Raise the World". In *Science Observed: Perspectives on the Social Study of Science*, edited by Karin Knorr-Cetina and Michael Mulkay (pp. 141–170). London: Sage.

Leinfelder, Reinhold. 2013. "Wir Weltgärtner". Interview with Reinhold Leinfelder. *Die Zeit*, 1 October.

Lovelock, James E. 1979. *Gaia: A New Look at Life on Earth*. New York: Oxford University Press.

Mukerji, Chandra. 2002. "Material Practices of Domination: Christian Humanism, the Built Environment, and Techniques of Western Power". *Theory and Society*, 31(1): 1–34.

Oxman, Neri. 2016. "Age of Entanglement". *Journal of Design and Science*, 1. Cambridge, MA: MIT Press.

Palsson, Gisli, Bronislaw Szerszynski, Sverker Sörlin, John Marks, Bernard Avril, Carole Crumley, Heide Hackmann, *et al.* 2013. "Reconceptualizing the 'Anthropos' in the Anthropocene: Integrating the Social Sciences and Humanities in Global Environmental Change Research". *Environmental Science and Policy*, 28(April): 3–13.

Plag, Hans-Peter. 2016. "The Laboratory for 'Patient Earth'". *Columns, On the Edge*, 8 July. http://apogeospatial.com/the-laboratory-for-patient-earth/.

Poel, Ibo van de and Lambèr Royakkers. 2011. *Ethics, Technology, and Engineering: An Introduction*. Hoboken, NJ: Wiley-Blackwell.

Rockström, Johan, Will Steffen, Kevin Noone, *et al.* 2009. "Planetary Boundaries: Exploring the Safe Operating Space for Humanity". *Ecology and Society*, 14(2): 32. www.ecologyandsociety.org/vol. 14/iss2/art32/.

Royal Horticultural Society, Media Centre (RHS). 2015. "RHS to Meet the Challenge of Climate Change and Help Gardeners be More Eco-friendly". http://press.rhs.org.uk/RHS-Science-and-Advice/RHS-to-Meet-the-Challenge-of-Climate-Change-and-He. aspx.

Schneider, Stephen H., Armin Rosencranz and John O. Niles, eds. 2002. *Climate Change Policy: A Survey*. Washington, DC: Island Press.

Schwarz, Astrid. 2016. "Kulturland in der Stadt: städtisches Gärtnern". In *Stadt und Land. Das Stadt-Land-Verhältnis und der Zwischenstadt-Diskurs im Spannungsfeld von status quo und utopischem Potential*, edited by Karsten Berr and Hans Friesen (pp. 181–196). Münster: Mentis.

Sloterdijk, Peter, Paul Crutzen, Mike Davis and Michael D. Mastrandrea. 2011. *Das Raumschiff Erde hat keinen Notausgang*. Frankfurt am Main: Suhrkamp.

Smetacek, Victor, Christine Klaas, Volker H. Strass, Philipp Assmy, Marina Montresor, Boris Cisewski and Nicolas Savoye. 2012. "Deep Carbon Export from a Southern Ocean Iron-fertilized Diatom Bloom". *Nature*, 487(19 July): 313–319.

Smith, Bruce D. and Melinda A. Zeder. 2013. "The Onset of the Anthropocene". *Anthropocene*, 4: 8–13.

Turner, Tom. 2004. *Garden History: Philosophy and Design 2000 BC–2000 AD*. London: Routledge.

UNESCO. 2018. Sustainable Development Goals. https://en.unesco.org/sdgs.

7 The distant gardener
Remote sensing of the planetary potager

Nina Wormbs and Johan Gärdebo

Introduction

The French word *potager* is old and means a vegetable garden, or, more precisely, a garden that could provide vegetables for a thick French soup, the potage. In the English context, the vegetables were destined for the pot. However, it is not just any vegetable garden filled with healthy, perfectly packed nutrients of different kinds, colours and shapes aimed at feeding family and friends. It may also be a *geometrically* organized piece of land, where the shifting seasons are visible and where mathematical beauty can take central position.

This chapter will explore resource mapping through the technology of remote sensing. We have chosen the potager, a specific version of the kitchen garden, as an introductory metaphor capturing the geometrical ordering of planetary resources. The planetary potager, or the planetary kitchen garden, thus conveys a resource dimension, including ideas on what we today call sustainability. In the early modern period this garden was sustainable, and in this chapter we will discuss how recent ideas on sustainability co-produced an understanding of the environment. The idea is not so much the particular size of this kitchen garden, even though we will discuss both size and scale, but more importantly *the way* in which it may be said to be gardened, or perhaps guarded. We will discuss the technology of remote sensing as a tool for sensing, assessing and mapping the resources in question and address how that information is used in actually appropriating, cultivating, harvesting and maintaining the land (Avignon, Dubois and Escudier 2014; Chuvieco 2008; Thompson 2007: 1, 11).

We suggest that the work done by remote sensing satellites is part of evaluating both the beauty and use of land – it is a form of gardening the planet. Our aim is to unveil the gardener as an actor, instead of treating it as a faceless, neutral and objective force whose ambitions and tasks are given and uncontested.

Furthermore, we propose that this activity may be understood as *environing* (Warde 2009, 2016; Nardizzi 2017), and that the technology used for this (i.e. remote sensing) is an *environing technology* (Sörlin and Wormbs 2018). The verb "environing" is an old English verb connected to the domestication of the land in and around smaller settlements and villages, and means to *make environment*.

Arguably, humankind has practically always environed, as the activity to make an environment for living is essential to life. Historically, much of that activity was carried out with small means, by hand, by horse and with the aid of different crafts. With the agrarian revolution, urbanization and the industrial revolution, the effects of these activities increased in scale. In the wake of the Anthropocene, environing uses technology to such an extent that an analysis of the outcome of the activity necessarily needs to include the tools. The technologies that might first come to mind are those that shape the environment when deployed. Trucks, drills, dynamite, fences, cement, etc. are technologies that shape and restructure, move and tear down, break and build. As outcomes, we can think of dams, deforestation, urbanization, mining, transport infrastructure or, for that matter, large-scale agriculture. However, an environing technology may also be something that works with our understanding of the environment. It may, for example, work on an epistemological level, structuring our thinking of limits and boundaries, challenging our understanding of the nature of a resource; it may work on an economic level, suggesting values and prices or demand and supply; or it may be aesthetic, picturing or imagining the landscape. In our case, the images from remote sensing satellites form a certain understanding of the Philippines; they environ and thereby structure the landscape as sustainable forests. Environing technologies contribute to forming our appreciation of what is possible to shape. A fishing quota guides the activity of fishing and thereby shapes the seascape by removing a certain amount of living stock. The idea of ecosystem services puts a value on a piece of environment and its use, and thereby channels economic resources to a specific and predetermined activity, disregarding another. In effect, environing is a complex system of activities that stretch from conceptual to material, depending on and supporting each other.

This chapter starts out with a discussion of remote sensing as an infrastructure and investigates the tension between integrity and increased information. It includes a section that deals with the development of remote sensing in the Western world. We then move to the empirical case in point, namely the mapping of land cover in the Philippines in 1987/1988, as an early example of a country-wide mapping by satellite for the purpose of sustainable development, financed by the World Bank and Swedish state interests. A discussion then follows that opens up the remote sensing images from being only about calculating material resources to their role in ascribing beauty, morals and political imperatives with regard to sustaining an environment, before we arrive at the conclusion.

Remote sensing as a resource infrastructure

Remote sensing does not have to take place from a satellite. For many years, aeroplanes were used to get an aerial view of the landscape and that technology is still important, as it has several advantages over a satellite. In this chapter, however, we will focus on remote sensing from satellites, as they allow for a planetary perspective, given their elevated position and global orbits.

When first imagined, remote sensing obviously gave rise to severe worry, as national integrity and sovereignty was at risk if the territory of one nation could be easily monitored by another. The idea itself had been around for many years, and earlier technologies, like rockets in the immediate post-war years, promised new possibilities, surpassing aeroplanes, which did violate national territories. However, the prospect took on a more material form following the launch of Sputnik in 1957. And as no diplomatic objection was raised in connection to the fly-over of this first artificial satellite, it in fact functioned as a precedent, establishing the principle of "right of overflight" (Launius 2014). A few years later, in 1959, the Committee on the Peaceful Uses of Outer Space was set up as part of the UN system, and with a Scientific and Technical as well as a Legal Subcommittee.[1] The "freedom of space", or the "open skies" doctrine, was also formally agreed upon in a resolution to the UN in 1961, after which a treaty was crafted.

At the same time, and not unconnected, the Cold War was a reality that made the two sides adopt new political positions, allowing for technoscientific development, which in turn fed back into the political process. The Sputnik chock enabled large US investment in science and technology, creating, for example, the National Aeronautics and Space Administration (NASA) and the Defence Advanced Research Project Agency (DARPA). A few years later, in 1961, the newly elected President Kennedy was faced with the Bay of Pigs invasion in Cuba and the spaceflight of Gagarin. It was in this context that he launched the Apollo programme (McDougall 1985; Muir-Harmony 2014).

In the Cold War technoscientific landscape, military goals could masquerade as scientific goals. The military funding of much of the earth system sciences is one example, which forcefully directed the course of USA research in the geosciences (Doel 2003; Oreskes and Krige 2014). However, military goals could also often be reached simultaneously with civilian goals, using the same technology. Meteorology is a case in point and the early US meteorological satellite TIROS, for example, served a dual purpose (Grevsmühl 2014; Edwards 2010: 219–222). Important also was the turn from using space technology to look outwards to instead look at the globe itself. NASA's *Mission to Planet Earth* was not only a quest for knowledge, but just as much a way to legitimize investment in this very expensive technology (Conway 2014).

Building on the notion of *information globalism* developed by Martin Hewson, Paul Edwards suggests the concept of *infrastructural globalism*, which he defines as the creation of "sociotechnical systems that produce knowledge about the whole world" (Edwards 2010: 25). It is not necessarily imbued with ideological beliefs but it is goal oriented, systematic and systemic with international epistemic communities as homes to central actors. The cross-border activities are crucial here, since they serve to establish encompassing technoscientific networks and systems. One such system is satellite remote sensing used to map large areas of the planet.

Edwards' focus is the global meteorological system for weather prediction with roots in nineteenth-century chartering of patterns and the establishment of synoptic measurements. However, his general arguments are just as valid also for remote sensing. The global meteorological system became three-dimensional with the advent of satellites and Edwards calls the process of making it so "technopolitics of altitude". Kennedy's suggestions on how to use space for peaceful benefit in the autumn of 1961 was, Edwards argues, a prime example (Edwards 2010: 215–223). We suggest that remote sensing is also an example of technopolitics of altitude.

There are several orbits in which remote sensing satellites can travel, but a common route is in a near-polar orbit c.800 km above the Earth's surface. After Sputnik, several scientific and surveillance satellites were launched in the late 1950s and early 1960s. Planning for remote sensing also began, but it took time before a satellite equipped with dedicated surveillance for remote sensing was launched. The first Landsat satellite, initially named Earth Resource Technology Satellite, pointing to the mission itself, was launched in 1972. Landsat was operated by NASA and built by Hughes Aircraft. The system is still in place today with eight satellites launched (2019). However, the challenges both before and during the operation of the system have been numerous. As Pamela Mack has shown, it was a classic case of technology push, where the American developers struggled to find uses and users (Mack 1990). In a study covering its use up until 2008, Brian Jirout has demonstrated the troubled path of organizing the Landsat system, first as state ownership, then as commercial enterprise, and most recently as a private–public partnership, as part of a longer struggle to find uses and legitimacy (Jirout 2016).

Vertical ambitions of a small nation: establishing Swedish remote sensing

Just like the US case, Swedish actors who wanted to sell remote sensing had a hard time. A scientific committee of remote sensing was established in 1969 under the aegis of the Board for Technology Development, the Remote Sensing Committee. When Swedish space activities were consolidated in 1972, the committee moved to the newly established Swedish Board for Space Activities. The Swedish Board for Space Activities was a small agency that to a large extent relied on the Swedish Space Corporation (SSC), a state-owned company created the same year to provide expertise on space technology in national as well as international settings. That the timing coincided with the restructuring of European space cooperation was no coincidence. Sweden had been part of the scientific collaborations within ESRO (staying out of the launcher organization for neutrality reasons) and when the new European Space Agency was formed in 1972 Sweden agreed to the "package deal". This meant membership fees and mandatory contributions. During these changes, the Swedish government also nationalized the sounding rocket facility Esrange, located in Sweden at 67 degrees north, not far from the mining town of Kiruna, which had been part of European cooperation during the 1960s (Wormbs and Källstrand 2007; Stiernstedt 2001; Hultqvist 2003).

During the 1970s, SSC aimed to not only continue but also expand the operations at Esrange, most notably by developing expertise and infrastructure for satellite remote sensing, with or without the support of the Remote Sensing Committee as it turned out. To that end the company began searching for applications for the technology. A number of projects were undertaken; for example, the mapping of oil spill in the Baltic Sea, monitoring forest damage in Scandinavia, and measuring water quality in Zambia.[2] However, the big cognitive and monetary shift came with the decision to join the French remote sensing satellite project SPOT (Systéme Pour l'Observation de la Terre), run by the French space agency CNES. Sweden declared its willingness to participate in the French project in 1977 at the ESA ministerial meeting and signed the agreement the following year. At the same time, SSC had secured its hold on Swedish remote sensing by establishing a receiving station for Landsat data at Esrange, and had been appointed the national point of contact for Earthnet, which was ESA's network for receiving, processing and distributing satellite remote sensing data.[3]

Part and parcel of the SPOT project was data management and image production. In both Sweden and France, SSC and CNES both established subsidiary companies, Satellitbild and Spot Image respectively, to produce and sell images from the remote sensing satellite SPOT. They expected users to be of many different types and there were high hopes for the future of remote sensing both in Stockholm and in Kiruna where Satellitbild had its main office, close to Esrange. However, following the launch of SPOT in February 1986, neither SSC nor its French colleagues had customers knocking on their doors asking for satellite images. The search for users continued (Borg 2017; Wormbs and Källstrand 2007).

Soon it was clear that these customers would not be found in Sweden but abroad. Through market surveys, SSC had concluded that one possible application for remote sensing would be as part of a system for Swedish aid to developing countries. In parallel, the Swedish government made various efforts to export Swedish expertise to other countries, partly as a means to alleviate structural challenges of Swedish industrial exports during the 1970s (SOU 1980: 23). As part of these efforts, the Swedish government founded the Swedish Commission for Technical Co-Operation (BITS) in 1979. Under the auspices of the Swedish Ministry of Foreign Affairs, BITS would promote Swedish aid and collaboration to developing countries where Sweden had commercial interests of some kind.[4] The initial projects expanded collaborations with countries where Swedish companies already had activities, but BITS also hoped to find new markets (Olofsson 2016; Klevby 2018). To this aim, BITS soon sought cooperation with the World Bank, which had financial troubles and needed additional partners, and was successful in establishing a few projects by the early 1980s.[5] In 1986, BITS and the World Bank set up a joint bank account, called "Swedish Consultants", to facilitate the flow of money between the two organizations.[6] These export projects also drew upon Swedish entrepreneurs working as aid consultants in Washington, DC. One consultancy, Swedish Projects

Incorporated, had initially formed as a subsidiary to the Swedish Chamber of Commerce in New York but was later privatized as a company specializing in supporting Swedish companies, like SSC, with exports to developing countries (Rasch 2016; Emsing 2016; Olovsson 2018).

Swedish Projects Inc formed part of an economic and organizational infrastructure, where state-funded companies sold their services to the third world. This "imperialism without imperialists" aimed to serve the national priorities of developing countries (Black in preparation; Magdoff 2003: 124–130). There are several examples of how Swedish engineers and scientists participated in development projects to build up new markets for Swedish expertise (Bruno 2016: 43–45; Öhman 2007: 18). It is important to note that Swedish foreign policy priorities of geopolitical neutrality and humanitarianism promoted, rather than restricted, such exports (Glover 2011: 101, 114–115, 194; Wiklund 2006). We will now describe how state companies, such as SSC, and private consulting firms, such as Swedish Projects Inc, used these Swedish foreign policy priorities in a project to map the Philippines.

Mapping the Philippines, 1987 to 1988

The Philippines became independent in 1946 after having been under Spanish, US and Japanese rule. President Ferdinand Marcos came to power in 1965 with the promise of ending corruption and bringing order to the land through a rule of experts. US aid, aeroplanes and satellites would provide an overview of the Philippines in order to redistribute land to the landless, which was the main cause for unrest. Instead, Marcos increased corruption, clung to power through martial law, and by 1986 faced a civil war, which could be averted by allowing a democratic election. Marcos lost to Corazon Aquino, who, like himself 20 years earlier, had campaigned on a promise to restore order, end corruption and conduct a land reform that gave rights to the landless peasants (Bresnan 1986: 256). Aquino vowed to be tough on the US government, since it had supported Marcos in exploiting the Philippines, but she also relied on international aid to reform the country, most importantly to swiftly finance and produce maps for conducting her land reform. One option would be to use satellite remote sensing from Landsat. However, many within the new government were sceptical of using American satellite systems, since this meant relying on US firms to map the country.

Through contacts at the World Bank, Swedish Projects Inc received news that the Bank had proposed a land cover map for "sustainable development" as part of supporting Aquino's land reform. Swedish Projects Inc thereafter proposed to the World Bank that SSC use French SPOT data for mapping the land cover of the Philippines. Since the Philippines were not among the countries to which Sweden prioritized aid, Swedish Projects Inc first secured funding from the World Bank before asking BITS to fund the rest of the project (Emsing 2016).

By using the services of Swedish Projects Inc, SSC was able to steer BITS and other Swedish interests towards supporting the Philippines. SSC saw Aquino's land reform as a golden opportunity to demonstrate the usefulness of satellite remote sensing. In a never-failing mood of optimism, topped with the self-assurance of those who know they are right and can see into the future, the project was described as the passing test for Swedish satellite remote sensing.[7]

The mapping of the Philippines would be a key to open doors for international aid projects in developing countries. In the long term, SSC aimed to use this international market to convince customers closer to home, who up until then had shown only modest interest in SSC's satellite data. By mapping the Philippines, SSC could demonstrate that it could produce real maps quickly and accurately, and central organizations like the Swedish National Land Survey might be forced to accept SSC's rightful place on a Nordic market.[8]

SSC, BITS and the World Bank continued to negotiate the extent and cost of the project well into March 1987. But in July 1986 SSC had already announced through its international newsletter that it would help BITS and the World Bank to map the whole of the Philippines.[9]

The World Bank and the Aquino Administration had agreed that a land cover map should be produced swiftly so as to support the land reform. SSC's map was in itself not meant for the redistribution of land but to provide information on what land resources currently existed, which allowed the reform to include plans for reforestation, farming systems, watershed protection and fishery rehabilitation.[10] Everyone agreed that equity and distributive justice had to be given to landless peasants in order to solve the root problem of environmental destruction. That SSC's land cover map served the sustainable development of the environment was meant as a support to the more important aim of implementing Aquino's land reform.[11]

Swedish Projects Inc mediated the negotiations during which a series of proposals were discussed on how extensively, in what detail and at what cost the Philippines should be sensed. If the study was to be conducted swiftly, it could not be too extensive or detailed in its analysis of the land cover. SSC had offered to either map the entire country or to provide detailed analysis for one part at a time, to make more use of the resolution in the SPOT data. Eventually the World Bank agreed to finance SSC to sense most parts of the Philippines and include a more detailed analysis of selected images, primarily as part of demonstrating the SPOT data's capacity as well as SSC's expertise in enhancing these data.

In order for Aquino to conduct the land reform, The World Bank stressed to SSC that the land cover map had to be finished within one year, to support an implementation of the reform in the spring of 1988 at the latest. This time constraint put limits on how many classes could be included when classifying land use in the Philippines.

There were a number of issues hindering SSC from sensing the Philippines. SSC had negotiated with the World Bank to secure more funding but in the process also delayed the project until March 1987. SSC worked intensively with gathering SPOT data in Esrange and to conduct fieldwork in the Philippines up until June when the cloudy wet season would hinder further work. Even under dry conditions, cloud cover often obfuscated large parts of the country's surface from SPOT to sense. SSC tried sensing parts of the country more than 30 times without obtaining sufficiently cloud-free images.[12]

In addition to SSC's constraints on sensing from above, the political situation in the Philippines also made sensing from the ground difficult. By the spring of 1987, opposition and violence had mounted from various groups. Peasants protested that reforms were too slow. The landowners, by contrast, claimed that the reforms were too large. The military, in turn, staged numerous coups, along with several attempts to take Aquino's life, under the pretext that national order could only be maintained by a more militaristic government. At one point the US Central Intelligence Agency intervened, without the consent of Aquino's administration, to strike against rebel groups. Several Filipinos and Americans were killed in the attacks and in subsequent retaliations (Wright and Bell 1988). In particular, the southern island of Mindanao was deemed too dangerous for international personnel to visit (Eriksson 2017).

SSC's fieldwork started from the north of the Philippines and progressed southward. The team consisted of both Swede and Filipino experts, each with somewhat different skills. The Filipino team members were land surveyors, trained in remote sensing technologies using aeroplanes. The material they gathered consisted mostly of aerial photographs, some dating back to the 1950s, and only covered parts of the country. The Swedish members consisted of engineers from Satellitbild, researchers from Swedish universities and development consultants.

The mapping work was conducted in three stages. First, SSC used SPOT to gather satellite data and print these as individual images. Second, the team brought the SPOT images to the Philippines for fieldwork to "ground truth" the images, which involved visiting places in the landscape that served as sample spots for larger areas. Ground truthing was sometimes conducted by aerial surveys, especially in the south where the situation was unstable and the government battled with rebel groups. Third, these two bodies of data were combined to determine which part of the satellite image corresponded to a certain class in the classification system. The team often consulted older data from Landsat or maps collected by the Filipino team members.[13]

When SSC began to sense the Philippines in March 1987, their plan was to map only 13 different classes for the map legend.[14] But as soon as fieldwork began in late April, this plan had to be revised to incorporate more information into the mapping. In the final legend, the number had increased to 24.[15] Most of these new classes came about as part of the fieldwork. The Filipino team members asked for revisions of land-use descriptions that corresponded to their own experiences, as well as to those of locals who the team interviewed while visiting the sample spots. The Swedish researchers demanded additional revisions on the basis that boundaries for the initial classification were too ambiguous. Not even when walking in the landscape, knee-deep in the grass of the sample spots, could the team members distinguish between some of the classes of land cover. Mapping this, the researchers claimed, could only be done by adding more classes (Alm 2017).[16]

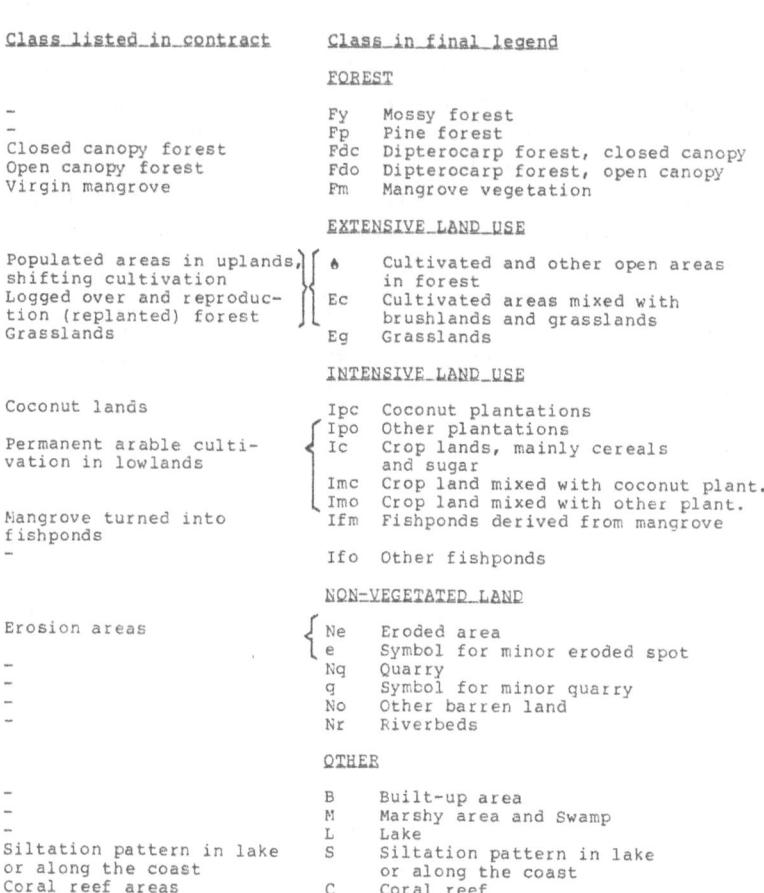

18

Class listed in contract	Class in final legend
	FOREST
–	Fy Mossy forest
–	Fp Pine forest
Closed canopy forest	Fdc Dipterocarp forest, closed canopy
Open canopy forest	Fdo Dipterocarp forest, open canopy
Virgin mangrove	Fm Mangrove vegetation
	EXTENSIVE LAND USE
Populated areas in uplands, shifting cultivation	▲ Cultivated and other open areas in forest
Logged over and reproduction (replanted) forest	Ec Cultivated areas mixed with brushlands and grasslands
Grasslands	Eg Grasslands
	INTENSIVE LAND USE
Coconut lands	Ipc Coconut plantations
	Ipo Other plantations
Permanent arable cultivation in lowlands	Ic Crop lands, mainly cereals and sugar
	Imc Crop land mixed with coconut plant.
	Imo Crop land mixed with other plant.
Mangrove turned into fishponds	Ifm Fishponds derived from mangrove
–	Ifo Other fishponds
	NON-VEGETATED LAND
Erosion areas	Ne Eroded area
	e Symbol for minor eroded spot
–	Nq Quarry
–	q Symbol for minor quarry
–	No Other barren land
–	Nr Riverbeds
	OTHER
–	B Built-up area
–	M Marshy area and swamp
–	L Lake
Siltation pattern in lake or along the coast	S Siltation pattern in lake or along the coast
Coral reef areas	C Coral reef

Figure 7.1 Illustration from final report for the Philippines project.
Credit: SSC.

The Swedish researchers asked for more details to be included in the classification, while the Filipino team members asked that these classes reflect customary uses and history of the land. SSC eventually agreed to negotiate a revised classification with the World Bank.[17]

The fieldwork meant an important change in the entire project. It reflected a more heterogeneous reality than was first expected, of which there are several examples from studies on the dilemmas of mapping (Rankin 2016: 2–16). Our concern, however, is not the accuracy with which the Philippines was mapped but with the use of the map. SSC revised its classification of the Philippine map to provide greater detail, in particular of forest classes. As we will see, this allowed the land cover map to be appropriated more readily for environmental purposes than had first been anticipated by Aquino's administration, the World Bank or SSC.

From land reform to land reforestation

Throughout 1987, in parallel to the mapping activities, a small change in the entire objective of SSC's project could be seen. Was it in fact possible to use the result to monitor deforestation and work towards its sustainability? The idea of deploying the project for sustainability had been formulated by the World Bank already in the early drafts for the mapping project,[18] but Aquino's main argument for swiftly mapping the country by satellite was to conduct a land reform (Säl 1987; Johansson 1987). The redistribution of land was the key to settling unrest in the Philippines. The Filipino government's failure to address the land reform issue was at the root of the country's continuing instability (Bresnan 1986; Hodgkinson 1988; Wright and Bell 1988: 36–37, 43).

After June 1987, when SSC had finished its fieldwork in the Philippines and began assembling the data in Kiruna, it first had to negotiate with the World Bank regarding a new classification for the map. The Bank accepted a changed, more detailed classification as long as this did not cause substantial delays in the delivery of the final map. The deadline had itself been postponed due to delays in Aquino's land reform.

SSC spent all summer, autumn and winter putting together the satellite data for interpretation and did so according to a specific procedure.[19] First, the Filipino surveyors provided official maps of the Philippines that would be scanned to create a raster. Second, SSC delineated regional and provincial borders from existing topographic maps. Third, the classes that SSC and the World Bank had agreed upon were divided into polygons that would be placed on the map to cover it as a layer. Fourth, the Swedish and Filipino team members worked with calculating the area size of each class province by province. Fifth, and finally, SSC summed up the map sheets to form statistical material, enumerating the extent of each class, for subsequent interpretations.[20] Areas that still lacked data, or contained land uses not included in the classification, were assimilated into other classes to avoid "white dots on the map" (Rasch 2016; Alm 2017).

As the political situation in the Philippines worsened, Aquino made requests for more help from the international community. In Sweden, a debate ensued focusing on the Philippines as an emerging democracy that Swedish aid should prioritize. It is against this background that one can understand the willingness of the Swedish government, and the Government of the Philippines, to refer to SSC's mapping project as a form of bilateral collaboration to support Aquino's land reform.

The deadline for delivery of the maps was 30 April and the team in Kiruna had to work hard to meet it. With only a few hours to spare, the maps were

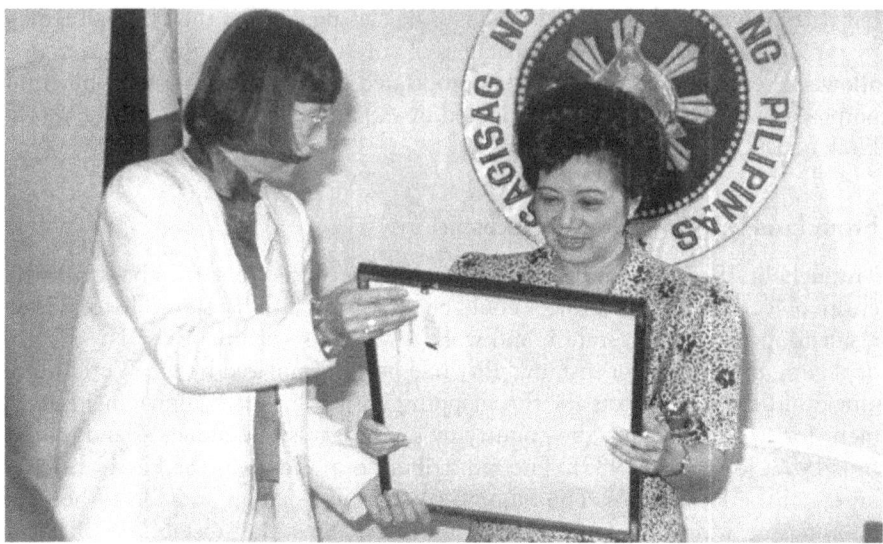

Figure 7.2 Hand-over ceremony, 2 May 1988.
Credit: SSC.

handed over to the World Bank in Washington, DC. At the meeting, the process of producing the maps was iterated and the CEO of SSC stressed in his speech the great potential of satellite remote sensing for environmental protection in developing countries (Lübeck 2014).[21] A few days later, on 2 May, a ceremonial hand-over was organized at the Swedish embassy in Manila, the capital of the Philippines. The Swedish Minister of Energy and Environment Birgitta Dahl personally presented the map to President Aquino (Dahl 2015). Local Filipino newspapers in Manila described SSC's map as "the first of its kind in the world". Furthermore, the map was said to provide vital input into Aquino's land reform, national policies on land use and the formulation of new legislation for conserving the rainforests.[22]

By the autumn of 1988, Aquino's administration professed that it was still pursuing a land reform. De facto, it had abandoned any attempts at implementing it. Aquino had decided that in order to remain in power, her administration had to balance powerful interests against each other. Political analysts at the time concluded that Aquino had postponed a comprehensive land reform since it risked provoking landlords into ousting the new Government of the Philippines (Hodgkinson 1988; Wright and Bell 1988).

In the years that followed, SSC conducted additional projects in the Philippines. Initially, it aimed to provide mapping for Aquino's land reform and also initiated negotiations on how to finance these projects with her government as well as with BITS.[23] However, instead of the land reform, Aquino's administration promoted SSC's land cover map as a demonstration of how the Philippines successfully used satellites to monitor, manage and sustain the country's forests. Aquino drew up a law against deforestation, with reference to the satellite map as a basis for defining, monitoring and regulating the boundaries of the forests. She also allocated resources to the Filipino land surveyors for continued use of the satellite data. In the following years, apart from passing legislation for reforestation, Aquino funded over 50 environmental projects based on SSC's maps and data.[24]

In 1990, the Government of the Philippines also organized conferences on sustainability where the results of these environmental projects, again with reference to SSC's mapping, were put on display.[25] The Swedish government participated and supported the Government of the Philippines in these efforts, stating, "This was the first time that a developing country had put so great weight on the term 'sustainable development'".[26] This turn towards sustainability was noted among both diplomats and remote sensing professionals in the years leading up to the United Nations Conference on Environment and Development held in Rio de Janeiro, Brazil, in 1992. The Brundtland Commission report was published in 1987 and the entire reformulation of the mapping project in the Philippines echoed the vocabulary of the new framework of sustainable development (Brundtland 1987).

Both the Government of the Philippines and the Swedish government heralded SSC's land cover map as a demonstration of how development aid could address environmental concerns. The project also formed the basis for continued bilateral bonds between Sweden and the Philippines. For many years to come, Filipino surveyors were invited to Sweden for training in environmental mapping in Kiruna. Correspondingly, SSC continued to carry out projects in the Philippines, as well as in various other places around Southeast Asia, always with reference to the Philippines mapping project (Bolton 2018; Emsing 2016; Engel 2016).

Discussion

The case of the Philippines environmental monitoring and mapping project illustrates how one political idea – in this case redistributing land for agriculture in a country ridden by inequality, instability and corruption – could be transformed into another political idea, namely the governing of global resources through the framework of sustainable development. The means for both the intended land reform and the resulting reforestation project was satellite remote sensing, a resource infrastructure. To some extent, the same transition could have taken place with another form of technology. Indeed, subsequent Swedish evaluations of SSC's project suggested that aerial reconnaissance was used as a supplementary technology when clouds obstructed sensing by satellite.[27] However, central to the actual realization of the project was that satellite remote sensing had very strong proponents. In fact, the primary reason for Swedish engineers to employ this technology in the Philippines in the late 1980s was to demonstrate that it could be done.

The Swedish state-owned company SSC had vertical ambitions that took the form of aid to developing countries. The first targets were close by, with monitoring of the Baltic. Second, projects in Eastern Africa were pursued, a region in which Swedish aid was particularly active. However, when the opportunity presented itself, Southeast Asia could work just as well to realize the vertical ambition. This technopolitics of altitude, where the consultants, engineers and surveyors in the US, Sweden and Philippines through employing remote sensing by satellite engaged in infrastructural globalism, had been successfully applied by the company before but with a totally different satellite application (Wormbs 2003).

SSC's land cover map of the Philippines served several interests. Swedish expertise could be exported to a new market in Southeast Asia, the World Bank could remain in the Philippines as part of supporting the new government, and Aquino's administration had access to satellite remote sensing as a tool for conducting its land reform. Aquino later used the detailed mapping of land cover to instead promote its use for environmental conservation, which did little for the landless peasants but helped maintain good relations with international organizations, governments and potential financiers for subsequent development aid to the Philippines. SSC's more detailed classification of land cover in this case also made it easier to apply for projects of environmental sustainability.

It was due to the lack of a sufficient land reform that SSC's land cover map gained increased importance as a contribution to environmental sustainability. Satellite sensing was not intrinsic to but instrumental of a shift towards environmental concerns. This gave the Aquino administration a tool not for land distribution but for land management at a time when it had to demonstrate the ability to govern the country without making enemies among its landowning elite. It was politically easier to address concerns about the environment than concerns about the ownership of land. Environmental aims became an alternative pursuit to legitimize the new government in the Philippines.

The interconnectedness of infrastructural globalism may at times be confused with equal opportunities for all. However, as Jason Beery reminds us, the occurrence of a global infrastructure does not necessarily mean that the sociotechnical system at hand creates a just and fair distribution of resources. Neither does terming something "global" mean that the resource is or should be equally available to everyone. On the contrary, he argues that the scale is not given, but is "produced through multiple social processes and strategies" (Berry 2016, 94). In the case of the Philippines, the revised legend illustrates forcefully how the combination of the view from above with local knowledge created a richer representation of the landscape (Nightingale 2003). However, it also points precisely to one of the central challenges with a global infrastructure like that of satellite remote sensing, namely the problem of scale. The resolution of the images was central to how they could be used. SPOT could observe things larger than 10 metres and did so in scenes that depicted a part of the Earth's surface 60 kilometres long and 60 kilometres wide. SPOT's resolution had been modelled to observe French farm fields, which were often smaller than those in other parts of Western Europe and the USA (Delclaux 2016). While the land reform needed detailed maps, it had been possible to derive sufficient scale from the data produced by SPOT as well as from civil satellites with less fine resolution, like Landsat.[28] But to see the trees and not only the forest, and to do so swiftly, one had to use satellites like SPOT. The land reforestation needed finer scales, whereas the land reform did not.

The resolution of SPOT also made SSC's project controversial for the US government which at this point had numerous military bases in the Philippines. In the spring of 1988, Swedish Projects Inc were pressured by US intelligence agents that satellite data, rather than maps, be delivered to the World Bank before being sent to the Philippines. This demand to review what military items could be seen in the data did not interfere with SSC's deadline for delivering maps over the Philippines, and therefore the Swedish consultants agreed to these demands (Rosenholm 2017; Emsing 2016). We do not know how this request would have been met, if indeed there had been problems with meeting the Filipino deadline. It is, however, an interesting way of exercising the neutrality that was purportedly an asset when receiving the commission in the first place.

Finally, the "imperialism without imperialists" that the mapping of the Philippines may be said to be was not aimed at resource extraction by a foreign power. The purpose of Swedish involvement was to be able to sell Swedish expertise, not Filipino forests. This is contrary to the US's use of Landsat for surveilling Sudan where the explicit goal was to be able to offer the US private capital information to extract oil (Black 2018). However, the resulting move from national land reform to sustainable development, arguably for the good of the entire planet, may be regarded as shifting the focus from local to global land use. How it eventually served local interests, which admittedly differed, is outside the scope of this chapter.

Conclusion

The French remote sensing satellite SPOT delivered a geometrically assembled Philippines, scenes that were perfect squares, 60 × 60 km in size. These were interpreted, categorized and transformed into maps that gave one version of the Philippines, a version that was a perfect fit to the present evolving discourse on sustainable development. The advent of satellite technology offered a new tool to cover and survey vast areas in a short amount of time, a tool to sense the Earth on a global scale. This technology created a Philippines of rainforests, part of a sustainable discourse on the global.

The core of the Anthropocene is the evident fact that humankind is changing the geology of the planet on a global scale. However, the concept also invites an epistemological as well as a conceptual discussion, side by side with the material dimensions of the new epoch. Satellites do affect our environment through their orbits and their debris, if we consider space to be part of our environment. We argue that the ways in which this technology helps the environment conceptually are just as important, de facto changing the view on the maps from one of land reform for national agriculture to one of beautiful, exotic and important rainforests of global interest. The technology allowed for a change from using the earth to using the Earth.

Agency in this story is dispersed and rests with the people in the Philippines, Washington, DC, Stockholm, Kiruna and Toulouse. The distant gardener is not easily spotted through gardening gloves or a pitchfork. However, the satellite in orbit, lacking both agency and morale, may still be useful as a metaphor for the fundamental shift in planetary appropriation, and fitting for the Anthropocene.

Notes

1 This committee has continued to grow and is today one of the larger UN committees with representatives from over 80 countries (www.unoosa.org/oosa/en/ourwork/copuos/history.html).
2 Stefan Zenker, Bidrag till historik över svensk Fjärranalys. 1986-02-23. Stefan Zenker, private archive, Sollentuna; SSC. Remote sensing. Information from the Swedish Space Corporation Newsletter, 2 May 1978. VBB JR-scannar i Zambia. P. 6; SSC archive, Solna.
3 The following builds on Johan Gärdebo's dissertation to be defended in April 2019. Some key references to a longer story are offered below.
4 The Swedish Commission for Technical Co-operation (SCTC/BITS). Cooperation between the World Bank and the Swedish Commission for Technical Co-operation (SCTC). To Frank Vibert, Office of the Senior Vice-President, The World Bank. 22 April 1983. Dnr U1167/83. Swedish Ministry of Foreign Affairs.
5 BITS. Bjerninger/Karlén. Bilaga ärende 7. 8 and 16 February 1983; Financial Times. "David Buchan examines the role of co-finance in Third World aid. World Bank seeks more partners". 1981-12-31; BITS. Bjerninger/Karlén. "Rapport. Besök på Världsbanken". 4 February 1983; BITS. P. Horm. Ärende 11. "Samarbete med Världsbanken". 19 April, 1983. National Archives of Sweden, Arninge. Swedish Commission for Technical Co-operation (BITS). U-krediter. Världsbanksgruppen. IBRD/IDA/IFC. 1980–1990. F2A: 73.
6 BITS. Dnr U1779. Doss 0.6.3. "Use of Swedish Consultants by IBRD and IDA". 4 June 1986.
7 For a similar project conducted by the SCC see Nina Wormbs, *Vem älskade Tele-X?: Konflikter om satelliter i Norden 1974–1989* (Hedemora: Gidlund, 2003), diss; SSC. IFA – 8. Lägesrapport för fjärranalysdivisionen, May 1987. C-G Borg. 1 June 1987, p. 8. SSC archive, Solna.
8 SSC. IFA – 8. Lägesrapport för fjärranalysdivisionen, May 1987. Claes-Göran Borg, 1 June 1987. SSC archive, Solna.
9 SSC. Remote sensing. Information from the Swedish Space Corporation Newsletter, 16 July 1987, p. 8. SSC archive, Solna.
10 John H. Cleave, "Philippines. Proposed Forestry, Fisheries & Agriculture Resources Management (ffARM) Study: A Strategy for Conservation and Protection". World Bank, 7 March 1986. National Archives of Sweden, Arninge. Swedish Commission for Technical Co-operation (BITS). Handlingar rörande Tekniskt Samarbete, TS, 1987–1993.
11 John H. Cleave, "Philippines. Initiating Memorandum. Forestry, Fisheries and Agriculture Resources Management (ffARM) Study: A Strategy for Sustainable Development and Conservation". World Bank, 15 September 1986, p. 8. National Archives of Sweden, Arninge. Swedish Commission for Technical Co-operation (BITS). Handlingar rörande Tekniskt Samarbete, TS, 1987–1993.

12 SSC, "Mapping of the Natural Conditions of the Philippines. Final Report". 30 April 1988, p. 5. SSC archive, Solna.
13 SSC. "Mapping", pp. 9–12. SSC archive, Solna.
14 SSC. Annex B. Project Proposal for Mapping of the Natural Conditions in the Philippines, 21 April 1987, p. 18. SSC archive, Solna.
15 SSC. "Mapping", p. 17. SSC archive, Solna.
16 SSC. Solna 17 August 1992/HCR. FILIPPINERNA. SSCs kommentarer till Scandinavian Project Managers' – SPM (Alm – Rylander) utvärderingsrapport av "Världsbanksprojektet" 1987/1988, p. 4. National Archives of Sweden, Arninge. Swedish Commission for Technical Co-operation (BITS). Handlingar rörande Tekniskt Samarbete, TS, 1987–1993. F2B: 208.
17 SSC. FGT11. "Mapping of the Natural Conditions of the Philippines Interim Report no. 1", 30 June 1987. SSC archive, Solna.
18 John H. Cleave, "Philippines. Proposed Forestry", 7 March 1986; "Philippines. Initiating Memorandum", 15 September 1986.
19 It was called the Scitex system and originated from an Israeli company that since the 1960s had produced systems and equipment for graphics design, printing and publishing.
20 SSC. "Mapping", pp. 12–13. SSC archive, Solna.
21 SSC. "Mapping", SSC archive, Solna.
22 Manila Bulletin, 1988-05-03, Tuesday. Satellite maps for CARP received. National Archives of Sweden, Arninge. Swedish Commission for Technical Co-operation (BITS). Handlingar rörande Tekniskt Samarbete, TS, 1987–1993. F2B: 208.
23 SSC. "Kartering av naturresurser i Filippinerna m.h.a. satellitbilder. Världens första vegetationskartering i stor skala med fjärranalys av satellitdata [Mapping of natural resources in the Philippines with the help of satellite images. The world's first vegetation mapping in large scale with remote sensing of satellite data]", 1 September 1988, p. 3. SSC archive, Solna.
24 The World Bank. From John Cleave, the World Bank, to Ingvar Karlén, Director, BITS, 25 May 1988; GoP Presidential executive order no. 192, section 22, 1987. In Evaluation of the satellite-based mapping of the natural resources of the Philippines byy Göran Alm and Lars Rylander. Final report, September 1992, p. 9. National Archives of Sweden, Arninge. Swedish Commission for Technical Co-operation (BITS). Handlingar rörande Tekniskt Samarbete, TS, 1987–1993. F2B:208; The Philippine Agenda for Sustained Growth and Development, July 1989. National Archives of Sweden, Arninge. Swedish Ministry for Foreign Affairs. BITS 1988–1992.
25 DENR. From Ricardo Bina, Deputy Administrator, NAMRIA, to Marika Fahlén and Carl Mellander, BITS. "Update on the Utilization of Products Provided by the World Bank-SSC Project on Mapping of Natural Conditions of the Philippines funded by BITS, Sweden in 1987–1988".
26 Swedish Ministry for Foreign Affairs. Embassy of Manila. Ch. Nilsson. No. 18, HP57. "Miljökonferens – Philippine Assistance Program (PAP)". Hans Grönwall, 1990-02-19, p. 2.
27 Evaluation of the satellite-based mapping of the natural resources of the Philippines by Göran Alm and Lars Rylander. Final report, September 1992, p. 8. National Archives of Sweden, Arninge. Swedish Commission for Technical Co-operation (BITS). Handlingar rörande Tekniskt Samarbete, TS, 1987–1993. F2B: 208; Interview with Göran Alm by Johan Gärdebo, 13 December 2017.
28 Hans Grönwall, Sveriges Ambassad, Manila. Pro memoria. 1988-05-24. Bilaga 1. Bo Eriksson. Bistånd till Filippinerna – en summering. 1988-05-15. p. 3. To kanslirådet Cederblad, Utrikesdepartementet (SE/RA/145/007). National Archives of Sweden, Arninge.

Bibliography

Alm, Göran. 2017. Interview by Johan Gärdebo, 13 December.

Avignon, Michel, Cathy Dubois and Philippe Escudier. 2014. *Observing the Earth from Space*. Paris: Dunod.

Berry, Jason. 2016. "Unearthing Global Natures: Outer Space and Scalar Politics". *Political Geography*, 55 (1 November): 92–101.

Black, Megan. 2018. *The Global Interior: Mineral Frontiers and American Power*. Harvard: Harvard University Press.

Bolton, Peter. 2018. Interview by Johan Gärdebo, 18 January.

Borg, Claes-Göran. 2017. Interview by Johan Gärdebo, 28 November.

Bresnan, John. 1986. *Crisis in the Philippines*, Princeton, NJ: Princeton University Press.

Brundtland, Gro Harlem. 1987. *Our Common Future*. Nairobi: United Nations Environment Programme.

Bruno, Karl. 2016. "Exporting Agrarian Expertise. Development Aid at the Swedish University of Agricultural Sciences and Its Predecessors 1950–2009". PhD dissertation, SLU Uppsala: Acta Universitatis agriculturae Sueciae.

Chuvieco, Emilio. 2008. *Earth Observation of Global Change. The Role of Satellite Remote Sensing in Monitoring the Global Environment*. Alcalá: Springer.

Conway, Erik M. 2014. "Bringing NASA Back to Earth: A Search for Relevance during the Cold War". In *Science and Technology in the Global Cold War*, edited by Naomi Oreskes and John Krige (pp. 251–272). Cambridge, MA: MIT Press.

Dahl, Birgitta. 2015. Interview by Johan Gärdebo, 26 May.

Delclaux, Philippe. 2016. Interview by Johan Gärdebo, 28 January.

Doel, Ronald E. 2003. "Constituting the Postwar Earth Sciences: The Military's Influence on the Environmental Sciences in the USA after 1945". *Social Studies of Science*, 33(5): 635–666.

Edwards, Paul N. 2010. *A Vast Machine: Computer Models, Climate Data, and the Politics of Global Warming*. Cambridge, MA: MIT Press.

Emsing, Erik. 2016. Interview by Johan Gärdebo, 30 November.

Engel, Pierre. 2016. Interview by Johan Gärdebo, 7 November.

Eriksson, Bo. 2017. Interview by Johan Gärdebo, 27 September.

Glover, Nikolas. 2011. *National Relations: Public Diplomacy, National Identity and the Swedish Institute, 1945–1970*. Lund: Nordic Academic Press.

Grevsmühl, Sebastian Vincent. 2014. "Serendipitous Outcomes in Space History: From Space Photography to Environmental Surveillance". In *The Surveillance Imperative: Geosciences during the Cold War and Beyond*, edited by Simone Turchetti and Peder Roberts (pp. 171–191). New York: Palgrave Macmillan.

Hodgkinson, Edith. 1988. *The Philippines to 1993. Making Up Lost Ground*. EIU Economic Prospects Series. London: Economist Publications.

Hultqvist, Bengt. 2003. *Space, Science and Me: Memoirs on Swedish Space Research during the Post-war Period*. Noordwijk: European Space Agency.

Jirout, Brian. 2016. "One Satellite for the World: The American Landsat Earth Observation Satellite in Use, 1953–2008". PhD dissertation, Georgia Institute of Technology.

Johansson, Folke. 1987. "Jordreformen i Filippinerna: Svensk bild kan hjälpa [Land Reform in the Philippines: Swedish Image Can Help]". *Uppsala Nya Tidning*, 16 October 16.

Klevby, Inga. 2018. Interview by Johan Gärdebo, 5 February.

Launius, Roger D. 2014. "Space Technology and the Rise of the US Surveillance State". In *The Surveillance Imperative: Geosciences during the Cold War and Beyond*, edited by Simone Turchetti and Peder Roberts (pp. 147–170). New York: Palgrave Macmillan.

Lübeck, Lennart. 2014. Interview by Lennart Björn, 10 December 2013 and 7 January 2014.

Mack, Pamela E. 1990. *Viewing the Earth: The Social Construction of the Landsat Satellite System*. Cambridge, MA: MIT Press.

Magdoff, Harry. 2003. *Imperialism without Colonies*. New York: Monthly Review Press.

McDougall, Walter A. 1985. *The Heavens and the Earth: A Political History of the Space Age*. New York: Basic Books.

Muir-Harmony, Teasel. 2014. *Project Apollo, Cold War Diplomacy and the American Framing of Global Interdependence*. PhD dissertation, MIT.

Nardizzi, Vin. 2017. "Environ". In *Veer Ecology: A Companion for Environmental Thinking*, edited by Jeffrey Jerome Cohen and Lowell Duckert (pp. 183–195). Minneapolis: University of Minnesota Press.

Nightingale, Andrea. 2003. "A Feminist in the Forest: Situated Knowledges and Mixing Methods in Natural Resource Management". *ACME: An International Journal for Critical Geographies*, 2(1): 77–90.

Öhman, May-Britt. 2007. "Taming Exotic Beauties. Swedish Hydropower Constructions in Tanzania in the Era of Development Assistance, 1960s–1990s". PhD dissertation, KTH Royal Institute of Technology.

Olofsson, Gunilla. 2016. Interview by Johan Gärdebo, 1 August.

Olovsson, Stigbjörn. 2018. Interview by Johan Gärdebo, 15 February.

Oreskes, Naomi and John Krige, eds. 2014. *Science and Technology in the Global Cold War*. Cambridge, MA: MIT Press.

Rankin, William. 2016. *After the Map: Cartography, Navigation and the Transformation of Territory in the Twentieth Century*. Chicago, IL: University of Chicago Press.

Rasch, Hans. 2016. Interview by Johan Gärdebo, 16 February.

Rosenholm, Dan. 2017. Interview by Johan Gärdebo, 20 November.

Säl, Ola. 1987. "Filippinska öar får svenskt bistånd via fransk satellit [Philippine Islands Receive Swedish Aid via French Satellite]". *Svenska Dagbladet*, 14 August.

Sörlin, Sverker and Nina Wormbs. 2018. "Environing Technologies: A Theory of Making Environments". *History and Technology*, 34(2): 101–125.

SOU. 1980: 23. Statligt kunnande till salu: Export av tjänster från myndigheter och bolag. Betänkande av konsultexportutredningen (Swedish Public Inquiry).

Stiernstedt, Jan. 2001. *Sweden in Space: Swedish Space Activities 1959–1972*. Noordwijk: European Space Agency.

Thompson, Kenneth P.A. 2007. *Political History of U.S. Commercial Remote Sensing, 1984–2007: Conflict, Collaboration and the Role of Knowledge in the High-tech World of Earth Observation Satellites*. PhD dissertation, Virginia Polytechnic Institute and State University.

Warde, Paul. 2009. "The Environmental History of Pre-industrial Agriculture in Europe". In *Nature's End: History and the Environment*, edited by Sverker Sörlin and Paul Warde (pp. 70–92). Basingstoke: Palgrave Macmillan.

Warde, Paul. 2016. "The Environment". In *Local Places, Global Processes*, edited by Peter Coates, David Moon and Paul Warde (pp. 32–46). Oxford: Windgather Press.

Wiklund, Martin. 2006. *I det modernas landskap: Kritiska berättelser och historisk orientering om det moderna Sverige 1960–1990*. Eslöv: Östlings Förlag Symposion.

Wormbs, Nina. 2003. *Vem älskade Tele-X? Konflikter om satelliter i Norden 1974–1989*. Hedemora: Gidlund.

Wormbs, Nina and Gustav Källstrand. 2007. *A Short History of Swedish Space Activities*. Noordwijk: European Space Agency.

Wright, Martin and Judith Bell. 1988. *Revolution in the Philippines?* Keesing's Special Report. Harlow: Longman.

8 Resistance in the garden

Nature and society in the Anthropocene

Davide Scarso

Introduction

The notion that human activities have attained a dramatic influence on life systems at a planetary scale is at the core of the proposal of the "Anthropocene" as a new geological age. For many, particularly in the human sciences, this entails a troubling blurring of the distinction between natural processes and the sphere of human relations, a dualism often deemed as one of the pillars of modern thinking (Chakrabarty 2009; Latour 2010). In order to be up to the challenges of the Anthropocene, or more in general of the profound changes in the functioning of climate regulations and other crucial Earth systems, the distinction between nature and the social should be reanalysed, critiqued and, eventually, disposed of. The proposal of the Anthropocene has met with several negative reactions and rebuttals (Wuerthner, Crist and Butler 2014; Suckling 2014; Haraway 2015; Hartley 2015; Moore 2016). In this chapter, we discuss not so much the refusal of the Anthropocene but more specifically the refusal of the need for or the opportunity of reassessing the difference between natural and social processes. While often interrelated, the two issues should not be conflated. One can believe that the opposition of nature and social life should be radically critiqued, at least in its mainstream incarnations, but should oppose or critique the Anthropocene proposal (Bonneuil and Fressoz 2016; Moore 2016). And, as arguably most Anthropocene scholars are from the natural sciences, one can be a strong supporter of the Anthropocene idea while maintaining, more or less explicitly, the general notion of a sphere of natural processes that are inherently, or ontologically, different from social affairs (or, to put it differently, that social affairs are inherently human).

We will then take into consideration two forms of refusing the reassessment or overcoming of the nature and culture divide that, in being particularly articulated and resonant, may be considered exemplary. They are also particularly interesting given their difference: apart from sharing a common dislike of the many contemporary theories that undermine the nature and society dualism, they could not seem more removed from each other. The first rebuttal is the "Half-Earth" proposal by US biologist Edward O. Wilson, fully developed in his 2016 eponymous book, while the second is the restoration of dialectical materialism by Swedish historian Andreas Malm in *The Progress of this Storm*.

Fifty per cent, or to each his own

Wilson, a sincere humanist and progressive thinker, argues the case for the protection of world biodiversity with the "Half-Earth Project", to which he dedicated several interventions (Wilson 2016b, 2017), a book (Wilson 2016a) and a website (www.half-earthproject.org/). The idea, in a nutshell, is this: "committing half of the planet's surface to nature" (Wilson 2017, 7). We need large areas of untrammelled nature in order to protect biodiversity, areas bigger than common parks and reserves. Perhaps more than 50 per cent is needed, but surely not less than that (plus, at least since Aristotle, halfway seems a reasonable compromise). In order to secure a liveable planet for future generations, in fact, at least 50 per cent of the whole of Earth must be dedicated to wild nature, or more exactly, to areas in which "biodiversity is to be returned to the baseline level of extinction that existed before the spread of humanity" (Wilson 2016a, 167).

Wilson, however, is not pointing at a nostalgic return to pre-modern techniques of social organization and to some sort of divinization of nature. On the contrary, in the other half of the world – "our" half so to speak – technological progress should be encouraged and increased, precisely in order to achieve "an authentic, predictive science of ecology". To attain a "sustainable Eden" we should not restrain from the plastic possibilities of synthetic biology, which is "the manufacture of organisms and part of organisms", in case you are asking. Only by promoting radical advances in biological research will reason finally triumph over superstition: "the goal is practicable because scientists, being scientists, live with one uncompromising mandate: Press discovery to the limit" (Wilson 2016a, 197).

To some extent, one may say that his proposal offers a salomonic solution to the decade-long struggle between biocentric and anthropocentric approaches to environmental issues.

Wilson's proposal literally cuts the Gordian knot in two, reserving half of the Earth's land and sea for a radically eco-centric approach while leaving the other half to the free development of human affairs: politics, history, science, technology, the arts, etc. – sort of unilateral, and hopefully final, truce after several centuries of human attacks on the natural world.

It should not be a surprise that for Wilson, the most insidious threat is not so much the globalization of neoliberal capitalism, for instance, or rampant industrialization or growing urbanization, but, rather, the argument according to which "pristine nature no longer exist, and true wilderness only survive as a figment of imagination" (Wilson 2016a, 74). In a chapter aptly entitled "The most dangerous worldview", Wilson stigmatizes the "new Anthropocene ideology", which by accepting that the natural world has become irremediably enmeshed with human affairs conveys a delusion of control while actually continuing to spiral towards self-destruction. Wilson mentions in particular Peter Kareiva, "a leading light of the 'new conservation' philosophy" and "leader of those who attack the existence of wilderness" (Wilson 2016a, 77).

In a way, Wilson's "Half-Earth" plan does not reject a form of eco-modernism provided that the "eco" part and the "modern" part are separated and that each has its own special area. What he finds troubling is the idea that ecosystems may be considered as potential objects of intervention according to specific human needs to be negotiated on a case-by-case consideration. As a good scientist, he recognizes that we do not really know everything about natural processes, that the "webs of life" are still in great part unknown, and that more research is needed. This is why he is not really calling for complete and sustainable management, since one cannot effectively manage something whose inner workings one does not fully understand. His point is to have areas as large as possible in which natural processes may occur as though no intentional intervention by human beings had ever taken place. Thus, he acknowledges that his "Half-Earth" plan will include some quite significant human intervention, but their exclusive objective is to undo previous more or less deliberate deviations brought about by the activities of human beings. Portraying this as management would introduce a discussion about goals and objectives, and perhaps even – heaven forbid – politics, which in the end is invariably human-centred.

Wilson's "Half-Earth" proposal results precisely from the refusal of a possible compromise between nature and culture, "a place where we can meet nature halfway", as Pollan would put it (1991/2003, 64). Nature is, by definition, that which is not human, human-made or human-influenced. Rather than the half-wild gardens scattered here and there (Marris 2011), the biologist, as we saw, believes we must arrive at restoring and protecting a wild half of the planet. The protection of biodiversity is as essential for humanity's survival as it is for life on Earth, so – apart from the necessary technical issues, for which further research is very much needed – there is no point in discussing options, choices or values: "To strive against odds on behalf of all of life would be humanity at its most noble" (Wilson 2016a, 4). And it would be difficult to find sound arguments to argue against the protection of "all life".

The Half-Earth project website, which informs us that American singer Paul Simon dedicated his last tour to fundraising for Wilson's initiative, is also asking its visitors, identified as "global citizens", to sign a "pledge". There is no information as to how many visitors have taken the vow. However, the online activist website Avaaz launched a similar petition in late 2016 to which, by the end of July 2018, more than 1.6 million individuals from all over the world had signed up.[1] The petition, which aims to achieve one million signatures and is getting a new one more or less every ten minutes, is directed at "world leaders", prompting them to "protect half of our planet". The explanatory text accompanying the campaign begins with a few paragraphs in quotation marks and in the first-person plural, supposedly exemplifying a direct speech from those sustaining the campaign. It reads as follows:

> We global citizens are deeply concerned by scientists warning that ecosystems critical to sustaining life on Earth could collapse in our lifetimes. We call on you to meet existing targets to protect biodiversity, forge a new agreement so that at least 50% of our lands and oceans are protected and restored, and ensure our planet is completely sustainably managed. This

must take into consideration the needs of human development, and have the active support of indigenous peoples. This long-term goal for nature can restore harmony with our home.

(Avaaz 2016)

The remaining text, without the quotes and not in italics, elaborates further: "By 2020 two thirds of wild animals will be gone", say the petitioners, because, with a quite vivid metaphor, "humanity is taking a chainsaw to the tree of life". However, not everything is lost, as "top scientists are backing an ambitious plan to put half of our planet under protection". Furthermore, "[scientists] say if we do it wisely, in a way that protects indigenous people from exploitation and land grabs, we can save 80–90% of all species!" (Avaaz 2016).

Although not explicitly mentioned, there is no doubt that "top scientist" Edward O. Wilson is the main source of inspiration for the Avaaz petition suggesting that half of our planet should be "put under protection". With its cursory homage to "indigenous people", to which we will return, it is worth noting that Avaaz's petition is structured as a discourse between "global citizens" and "world leaders", without any political context or references to any sort of mediating institution. It is a huge number of individuals addressing their "leaders", whoever and wherever they are, and the only thing these "global citizens" have in common, apart from being human earthlings, is that they have signed the petition and, therefore, share its content. But the "post-democratic" character of this discourse does not end here. The aim of the petition is not so much to put pressure on "world leaders", but rather to "make this solution so famous, our leaders can't ignore it" (Avaaz 2016). That is to say, "world leaders" should adopt the solution not because many people are asking them to do so but because, once they have had the chance to know it, they could not possibly fail to see that, being formulated by "scientists" (even better, "top scientists"), it is inherently good for everybody. The matter of how and by whom it may come to be implemented is almost irrelevant.

Wilson's Half-Earth proposal shares in large part this "post-democratic" attitude. As a matter of fact, the term "politics" never appears in the book, while nine times out of ten the adjective "political" occurs as part of the expression "political leaders". According to the biologist, a "sustainable Half-Earth system" would have its natural institutional foundation in the World Heritage Fund of UNESCO. As we know, by the end of 2017 President Trump had withdrawn the USA from UNESCO based on an alleged anti-Israel bias (Ronald Reagan did the same in the 1980s because of its alleged pro-soviet bias). While in both cases complex networks of national and international interests were crucial, both depicted their decision as a reaction to the political non-neutrality of UNESCO's actions. But the fact is precisely that protection and conservation, for cultural but also for natural sites, will seldom be socially and politically neutral (Silberman 2013). For Wilson, as for many others, in making the distinction between natural phenomena and the social world as a backdrop, environmentalism is inherently non-partisan and its actions not only can but should transcend politics and democratic practices. Environmental actions are guided by science and technology, and no other engagements have such an alleged "rational and neutral basis" (Armiero 2015). Michael Pollan, who, already in 1991, argued that the notion of wild or pristine nature was acting as a "taboo" and a "profoundly alienating idea" struck a chord when he acknowledged that it often acted as "a check on our inclination to dominate and spoil nature" (Pollan 1991/2003, 214). This is precisely why many conservation biologists and deep-ecologists share a "biocentric" approach to environmentalism. The inherent, "sacred" value of nature lies precisely in it not being human, and just as human affairs are contingent and, in the end, always open to discussion, so nature is entirely in itself, self-contained and amoral.

The "Protect Half of Our Earth" petition mentions the need to "have the active support of indigenous peoples" (Avaaz 2016), a reference that was probably due to the absolute lack of any reference to it in Wilson's discourse. Native people are mentioned only once in Wilson's book, when he observes that "wildernesses have often contained sparse populations of people, especially those indigenous for centuries or millennia, without losing their essential character" (Wilson 2016a, 77–78). Those "sparse populations" are defined precisely by the fact that they somehow manage to live a human life without depriving the wilderness of its essence. Critics of the "Half-Earth" proposal highlighted that in the case of Mozambique's Gorongosa Park, for example, pointed out as a model of successful conservation by Wilson, the strong opposition from local communities was completely disregarded (Büscher and Fletcher 2016). Considering that "most existing 'wilderness' parks have required the removal or severe restriction of human beings within their bounds" (Büscher and Fletcher 2016), the lack of any reference to the social, political (and even military) implications of the "Half-Earth" plan is remarkable (Dowie 2005, 2009).

From this point of view, the "rambunctious gardens" of the Anthropocene seem to offer more space to a political discussion that takes into account the entanglements of ecosystems and social formations. If there is no objective and undisputable notion of what nature is and different natural areas can be dedicated to different ends, at least some debate and negotiation will be needed in order to assess this. As Marris (2011, 170) puts it: "*Society must decide* what its goals are on multiple scales, then allocate the best-suited land to these *various goals and get going*, not shying away from the occasional bold experiment." Framing environmental issues in terms of options and goals seems to open up more space to political confrontation than referring to humankind or life as such. This is what political ecologists have shown. But when we say that "society must decide" or that priority should be given to the protection of ecosystems "that benefit us", we are implying that social relations forms an organic unity in which decisions are made for the benefit of all of its members, and that the "we" of "society" is a rather straightforward and generalizable notion. This, however, is unproblematic only within the limits of a – rather naïve– deliberative concept of politics, in which all different participants are duly represented and all possible options are clearly stated and accessible.[2]

Smashing the framework

To sum up the situation, the climate is warming because many people have burnt and keep burning a lot of fossil fuels which brought atmospheric CO_2 to levels that, given its greenhouse effect, are incompatible with the relatively stable climate of the Holocene. The climate does not change in a linear fashion and several feedback effects may bring about abrupt changes in climate regime, and possibly even lead to runaway global warming (Goldblatt and Watson 2012). A gradual phase-out of fossil fuels on a global scale seems presently either unlikely or much too slow; thus, according to Swedish historian Andreas Malm, the only way to avoid warming levels that would be catastrophic for human societies is immediate direct action, i.e. "smashing the fossil infrastructure" (Malm 2018, 155). The contemporary arguments for overcoming or at least blurring the distinction between nature and society (social constructionism, actor network theory, new materialism), despite their fashionable radical allure, are actually hindering the conditions that allow or instigate direct action. These conditions, Malm argues, are on the contrary maximized by Marxist dialectical materialism, which rests on a clear-cut distinction between nature and society considered as opposite poles in a reciprocal relationship: "As William Petty says, labour is the father of material wealth, the earth is its mother" (Marx, referenced in Malm 2018, 159). The different brands of "hybridism", arguing that the distinction between the sphere of natural phenomena and that of social relations has become blurred, or never actually existed, are either reducing the former to the latter, as do social constructionists, or are depriving the social or human pole of its defining attribute, that is, agency. These would be playful intellectual vagaries were they not applied to climate change, as their proponents stubbornly insist on doing. A "hybridist" approach to climate change would mean either that global warming, being socially constructed, can be simply evacuated by a conceptual effort or that, as natural processes have their own agency, there is no point in fighting or even opposing it (see Malm (2018, 149–156) for a succinct recapitulation of the general argument).

Malm is right in pointing out the fact that often professing the end of the distinction between nature and society, or the need to overcome it, has become an intellectual cliché and an empty gesture (see Ingold 1996; Scarso 2013). In the human sciences, it generally means simply that nature has been absorbed by an all-encompassing sphere of social meanings and interactions, and vice versa, in the natural sciences.

He argues, however, that any hesitation as to the universal and trans-historical character of the distinction between nature and society necessarily entails a position in which any difference among the entities that populate reality is elided, in what some call a "flat ontology". Arguing that the characteristics of the elements that compose a given state of things cannot be judged aprioristically, and without taking into account the very state of things in which they are involved is subtly but radically different from believing that those components are precisely equivalent to the point of being indistinguishable. To say that two things are not *necessarily* different does not mean those two things are *necessarily* the same.

To show how something may result from the combination of elements pertaining to different categories without entailing that those categories are thus rendered irrelevant or nonexistent, Malm gives the example of the Druze faith, "in which doctrines of Hindu, Shi'ite, Platonic, Gnostic, Christian, Pythagorean, Jewish and other provenances are drawn together" (Malm 2018, 48). By the fact that all those different sources are incorporated into a new synthesis, a scholar of the Druze faith would not feel compelled to dispose of the categories of Platonism, Shi'ism, Gnosticism, etc. But what this example makes clear is that Malm does not believe that the Druze "belief system" has the same originality and autonomy – let us say the same ontological status – as that of the sources it brings together. Platonism and Shi'ism are apparently well-defined and self-sustained cores of meaning that do not derive their existence from outer sources; they subsist by themselves or, in other words, are substances. Not so for the Druze faith, which is essentially derivative; however novel the combination it produces, it does not constitute a substance and is thus necessarily accidental. Perhaps this is due to the fact that the Druze faith came *after* Christianity, Platonism, Shi'ism, etc., and it is thus a subsequent patchwork of different traditions. But – and lacking any special knowledge on the issue I am speculating about – perhaps one should not exclude beforehand the hypothesis that the Druze system revealed some affinities and reciprocal influences among its sources, or maybe even common theological nuclei that were around before they became formalized in one or the other of those traditions. Speculations aside, for Malm there are things that belong entirely to certain given categories and then these things may combine and give way to new composite entities. He admits that the combinations of natural and social elements could very well be called hybrids, had not the term already been hijacked by Latour and his acolytes and with certain caveats. Malm's combinations could indeed be labelled hybrids, provided that the hybrids' existence did not call into question the elements that composed it, as apparently contemporary "hybridists" argue, just as a mule does not disprove that donkeys and horses are different species (this would entail excluding hybrid speciation and, more in general, assuming that the concept of species is relatively straightforward). In particular, most of what exist can be divided into two reciprocally exclusive main groups, namely nature and society, everything else being the result of the interaction among different elements belonging to one of those two groups. Cartesian dualism, however, the pet hate of postmodern critique and of ecological thinking, looms large. While the author agrees that Descartes' ontological dualism may be the most relevant root of the illusion of radical detachment between human affairs and the surrounding natural phenomena, he argues that the refusal of Cartesianism does not necessarily have to result in ditching any form of distinction between mind and matter.

According to Malm, by eliding the difference between mental phenomena and material substrata or, more broadly, between nature and society, one would miss the specificity of human agency and surrender to political powerlessness.

This is why he adopts "substance monist property dualism" (Malm 2018, 53), a subtle metaphysical position in which, while acknowledging the common belonging of all that there is to one and the same substance (i.e. matter), it also acknowledges that some of what there is may be endowed with properties not found elsewhere. Reviving a long tradition of human exceptionalism, which has somewhat fallen into disgrace in recent years, Malm defends that human beings possess the unique and distinctive capability of mental experience and that this sets them apart from all other natural phenomena. Although the mind is not some mysterious gift from God bestowed upon them but an emergent property of certain organic processes, it is because they have a mind and can thus intentionally shape their actions that human beings act upon nature. Humankind then entertains a paradoxical relationship with nature, as that to which it belongs – human beings do have organic bodies and live in a material world – but at the same time "that from which it seems excluded in the very moment in which it reflects upon either its otherness or its belonging" (Soper, cited in Malm 2018, 67). In order to trace what he considers to be a neat distinction between natural phenomena and human actions, and through an extended review of recent debates in the area of philosophy of mind, Malm adopts quite vehemently the definition proposed by Kate Soper: nature is

> [T]hose material structures and processes that are independent of human activity (in the sense that they are not a humanly created product), and whose forces and causal powers are the necessary conditions of every human practice, and determine the possible forms it can take.
>
> (Malm 2018, 28)

This is why climate change, for instance, is not a social construction. If to construct something is "to inaugurate a product which previously did not exist" (Soper, cited in Malm 2018, 38), it is quite self-evident that no human being participated in the inauguration of the Gulf Stream or created CO_2 so that it traps heat in the atmosphere. By burning enormous quantities of fossil fuels some human beings influenced and altered these mechanisms, but they did not create them. In other words, certain natural processes have been radically altered by some other and quite distinct social processes, but they have not been constructed or created.

While apparently quite straightforward, the distinction is not without its problems. The human body, for instance, is by all means a natural entity, often regarded precisely as the most eminent proof of the fact that human beings belong to nature. It could be considered a "social construct" only in its cultural and symbolic interpretation, not in its "functioning". But take human bipedalism. The capacity of moving by means of two legs is an essential feature of human beings and had a crucial impact on human evolution. Bipedalism, however, is not exactly an innate, or "natural", feature. As a dramatic case of early neglect in orphan children showed (Tardieu 2012), infants who are not guided and supported by a significant caregiver in their attempt to walk upright do not develop bipedal locomotion. And if they do not "learn" to walk at a certain sensitive age, their skeletons will develop in such a way as to make upright walking physically impossible (Tardieu 2012). This means that the modern human body, in order to fully develop, needs the presence of child rearing, and thus of a social life, however minimal. One could argue, therefore, that some form of social behaviour pre-existed, or co-existed with, the development of a phenomenon considered natural, namely the current average structure and functioning of the human body.[3] This is to say that labels such as "society" and "social relations" may not necessarily constitute a coherent and well-delineated category of phenomena that somehow emerged from natural processes a few millennia ago and has since then maintained its unaltered general properties through innumerable different manifestations. Quite surprisingly for someone so well read in what he calls "the Marxist canon", terms like "society" and "social" (as well as "human", "mind", "history", etc.) are somehow exempt from any form of ideological implication or historical consideration. Primitive human groups probably lived "fettered to the moment and its pleasure and displeasure", as Nietzsche says about cattle in the *Untimely Meditations*, but since they left their prehistorical and pre-social bliss, their social and historical beings have become their permanent and inherent feature. While paying lip-service to historical consciousness, because everybody knows that social configurations do change, the very category of the social is to some extent "naturalized". On the contrary, far from being a neutral and objective concept, the very notion of an autonomous and self-contained sphere of "social relations" is deeply rooted in the genealogy of modernity, and served, and continues to serve, specific ideological functions, which Malm's binarism renders impossible to evaluate critically.

Setting aside the metaphysical subtleties of property dualism,[4] the naturalization of the social underpins much of Malm's intransigent and negative militantism. As we saw, hybridism implies the dissolution of any distinction between nature and society, and this is bad because it blurs what one can or cannot do. If one sets out to draw a theory that may encourage and sustain militant action, one should recognize that "Nature is real; nature and society form a unity of opposites; society is constructed" (Malm 2018, 156). Following Kate Soper, nature is that which is independent from human action and that is not a human product. This is why natural processes are unalterable and one cannot change the fact that water freezes at zero degrees Celsius, for instance. Society is pretty much all of the rest; that is, all that which is dependent on human action and is by its essence the object of human intervention. Natural substrata dictate several general guidelines, which leave however plenty of space for human beings to shape their collective lives in many different ways. Climate change is precisely the result of a specific historical formation, capitalism, and more specifically the fossil economy, having crossed a few of those guidelines and upset the mechanisms of climate regulation that characterized the Holocene. Summing up his position, Malm (2018, 156) writes: "agency cannot be found in inanimate matter but may still appear among human collectives, which can potentially target the incumbent technology that embodies social power – these are some of the necessary premises for an activist theory".

Climate change is already threatening the subsistence of the poorest communities at the periphery of capitalist affluence and, in the longer term, the very existence of complex societies, so Malm's resolute call to direct action is an honourable stance. The point is that, despite his frequent reference to the "Marxist canon" and his persistent "Marxist parlance", his argument does not sound particularly Marxist, much less Marxist dialectical materialism. Believing that capitalism fosters inequality and is based on a prospect of continuous economic growth that is incompatible with life in a finite world, saying that a radical change of paradigm is urgently needed and manifesting a lack of faith in the transformative potential of representative democracy do not guarantee, by themselves, a certificate of Marxism. The fact that such beliefs are shared by people as different as George Monbiot (2018), Paul Kingsnorth (2013) and Mayer Hillman (Barkham 2018) is telling, as is the fact that Malm indicates Naomi Klein's *This Changes Everything* and Carolyn Merchant's *Death of Nature* as essential references in the ecological Marxist canon. These two books are inescapable contributions, respectively, to climate activism and to the history of the attitudes towards nature and gender that stand by themselves and to which Malm's Marxist stamp of approval does not provide any added value. Anticapitalist and ecological activism are by all means honourable ethical positions but, unless accompanied by an analysis of the relations between the forces at play and the evidence that their structure and contradictions are creating the conditions for the overturning of those relations into a society liberated from human and natural exploitation, they do not seem to have any specific "Marxist import". Calls to "smash [...] the framework" or to "expropriate the 1%" building on rage and panic for a warming climate, in which a "fringe of more or less deviant personality types ready to *act*" (Malm 2018, 137, emphasis in original) will one day "combine" with the masses expropriated by global warming, with no further qualifications, is pure voluntarism. To leverage social unrest caused by climate disaster for a revolutionary intervention against the fossil economy as an opportunity to deploy "a conscious programme aimed at creating or remodeling whole social structures" (Anderson, cited in Malm 2018, 118), out of pure goodwill is – here literally – unbridled social constructionism "in Marxist garb". Even if one concedes that "[o]bjectively speaking [...] the liberation of nature is a global class demand" (Malm 2018, 208), unless the class of those who are expropriated by climate change are the carriers of the values of the coming liberated society (and not only "objectively"), revolutionary activism is nothing more than voluntarism.

It it also interesting that Malm (2018, 227) concedes that, if things turn uglier than expected from the point of view of climate, "a detour of fighting for a planned phase-out of solar radiation management" could constitute part of the "revolutionary project for the next few centuries or so". There is no mention as to what powers, forces, institutions or social arrangements may implement a "solar radiation management" (which is geoengineering, in case you are asking), for several centuries and supposedly on a global scale, and who and why would then plan its "phase-out". Admittedly, a world without fossil fuels and climate warming would be better than this one, but nothing tells us it would be inherently non-capitalistic, less unfair or more ecologically minded. Unless one considers, as Malm does in a reference to Klein, that "[t]he power of the sun, wind, and waves can be harnessed, to be sure, but unlike fossil fuels, those forces can never be *fully possessed*" (Klein, cited in Malm 2018, 228, emphasis added). Thus, given their "inappropriable" character, shifting to renewable sources would per se imply a change in power relations between nature and human society (and perhaps, who knows, also between classes). Considering, however, how hydroelectric power has been a staple of nationalism and imperialism (Pritchard 2012), for instance, and that there is no reason to believe that capitalism is *inherently* incompatible with renewable energies, this sounds like well-intentioned wishful thinking rather than "Marxist parlance".

Conclusions

As we have seen, Andreas Malm believes that blurring the frontier between nature and society would be fatal for the future of humankind. While he admits that this does not mean "of course that a warming planet can be literally cut in two halves", he argues that "the *analysis* of it must execute a similar operation" (Malm 2018, 75–76, emphasis in original). Even though, apparently, on a radically different plane, Wilson also feels the need to clarify that his solution "does not mean dividing the planet into hemispheric halves" (Wilson 2016a). Despite their differences, they share the notion that, though broad categories such as these are difficult to pinpoint, nature and society are nonetheless defined by their mutual exclusion, an exclusion that should be upheld, as we saw, analytically for Malm and quite literally for Wilson. For both, it is natural that which is not human just as it is specifically human that which is not natural, and we should keep it that way. Furthermore, precisely because of its inhuman character, nature is – again, by definition – an independent, self-sufficient and non-contingent sphere, it is made up of mere facts and leaves no room for choices and values. Society, on the contrary, is essentially multi-form, mutable and, therefore, historical and political.

Along with these basic assumptions, Wilson and Malm share also a rather patronizing and dismissive attitude towards indigenous groups. We have seen that the Half-Earth project basically ignores the fact that its implementation would entail the removal and forced transference of many communities now living in the areas to be "put under protection". The "non-natural" half of the world would probably have to take in millions of "conservation refugees".[5] Environmental journalist Jeremy Hance (2016) quotes Wilson as saying that native communities "are often the best protectors" of their own lands; thus, according to him, reserving half of the planet to nature "would not simply mean banning people from half of the planet's land area, but keeping these areas undeveloped". This means that he ignores, or perhaps downplays, the environmental import of indigenous practice like slash-and-burn agriculture, hunting, or even just tending gardens or involuntarily dispersing certain seeds (on this, see Kawa 2016). But this also means that if an indigenous person had the terrible idea of taking up a gasoline chainsaw, a rifle or a smartphone, he or she would instantaneously cease to be indigenous, at least in Wilson's eyes. This makes perfect sense once you consider that, for Wilson, indigenous people are those human groups that somehow manage to inhabit wild nature without spoiling its very wildness. Again, real nature is the non-human, which is why, according to the Half-Earth project, in the 50 per cent of the Earth's surface to be devoted to nature, "biodiversity is to be returned to the baseline level of extinction that existed *before the spread of humanity*" (Wilson 2016a, 167, emphasis added).

The dialectical character of the relation between nature and society, according to Malm's argument, is equally hinged on a mutual definition of the two poles; one is what the other is not, like – to use a somewhat worn-out metaphor – black and white. And we find an equally dismissive attitude towards indigenous groups. Perhaps "cultures" where "no boundaries are drawn between the social and the natural" have been common in human history, but this "hardly ratifies them" (Malm 2018, 173). Embracing cultural diversity as such would lead "down the slope where everything and nothing is true and false at the same time" (Malm 2018, 173). The social pole is intrinsically subjective and polymorphous; there is nothing that is not contingent, except the fact that it is not natural. The distinction between nature and society itself is not cultural, but transcendentally true, so to speak. So, indigenous people may carry on with their "animist ontologies", which can eventually represent a sort of "fellow traveller", but the real commitment should be "revolutionary ecological practice" (Malm 2018, 174). Again, the notions of distributed and non-human agency are often invoked as empty slogans, and, even more to the point, what Malm calls "double monism" and flat ontology really run the risk of depoliticizing social and environmental issues. I believe, however, that the most interesting thinkers in the Anthropocene debate, first and foremost Bruno Latour, are not aiming at a flat ontology, a double monism or, even less, social constructionism. The critique of "the Great Divide" (Latour 1993) between nature and society is aimed at multiplying differences, not eliding them. The entities that correspond to the traditional categories of nature and society do not disappear suddenly, nor are they simply blended into a formless goo, but may be decomposed and redistributed according to different, and contextual, criteria (Latour 2014). During his time among the Ojibwa people in northern Canada in the 1940s, the North American anthropologist Alfred I. Hallowell questioned an elder informant about his "animist" beliefs: "Are *all* the stones we see about us here alive?" After lengthy reflection the elder man replied, "No! But *some* are" (Hallowell, cited in Ingold 2000). I believe this is a good way of saying that agency should not be considered as the permanent attribute of a certain class of entities, be they human or not, but as a relational or contextual property of certain entities as they enter the composition of certain assemblages. As historian Timothy

Mitchell would say, agency is always hybrid, which does not mean that it is undifferentiated and more or less equally distributed. On the contrary, human beings always tap into non-human forces and energies in order to pursue their goals, and it is not any goal that is allowed by the force they tap: "So-called human agency draws its force by attempting to divert or attach itself to other kinds of energy or logic" (Mitchell 2002, 29).

I believe that critical thinking, coming from any disciplinary area, should attend to the strange partnerships between humans and non-humans that have constantly been made and rearranged through history. In particular, we should follow closely all the situations in which this partnership is being renegotiated, where its present forms are in crisis and where maybe some of its new forms are being resisted and opposed. This is where indigenous people come in. Not because they happen to live in that other half of the planet which is more green than ours, so they should be asked permission, but because their fight for their land and for their relationship with their land represent an extreme point of resistance to capitalist development, perhaps to modernity as such, that force us to call into question many of our political and environmental assumptions. Member of indigenous groups risk their lives every day while they protect their land from loggers and developers. They also fight for representation, for their rights to autonomy and self-determination to be acknowledged, for their demands to be represented by the government institutions that administer the territories in which they live. But one could say that theirs is not only an inescapable political and environmental struggle but a form of cosmopolitical resistance. Amerindian thinking is not merely an example of a different "representation" of nature, but a difference that resists representation itself, just as their fight is a form of "irrepresentable" resistance. There are many places in which the relationship between nature and the social world is being reassessed and reworked, with different implications and different intensities, often calling directly into question the powerful effects of scientific expertise and modern political categories. Think of any conservation or re-wilding project, local communities fighting water, air or land pollution in their area, even something a little crazy like resistance to obligatory human vaccines. Critical thinkers should follow closely all these shifting and impure points of friction, and participate in the creation of platforms and concepts that may allow us to properly analyse and comprehend their implications, because it is there that new political perspectives, and new forms of collective freedom and emancipation are possibly being opened.

Acknowledgements

FCT MCTES – Project PTDC/IVC-HFC/6789/2014 – ANTHROPOLANDS – *Engineering the Anthropocene: The Role of Colonial Science, Technology and Medicine on Changing of the African landscape.*
FCT MCTES – Project PEst-OE/HIS/UI0286/2014.

Notes

1 Avaaz's petition web page is not dated; however, the oldest records available through a web search go back to November 2016.
2 The Breakthrough Journal shows a growing concern with the political implications of environmentalism, to the point of declaring that "nature is political" and that "the impulse to naturalize [...] is a power-play" (Brush and Nordhaus 2018). Nevertheless, even the issue on "Democracy in the Anthropocene" (2017) brings little light in this regard.
3 The same argument could be extended to other characteristics of the human body such as the size of the head, and related difficulties at birth, or prolonged infancy.
4 It is particularly difficult to follow Malm when, after arguing for several pages that human behaviour has a material substratum but is set apart from nature because of its intentional or mental character, he concludes that the whole issue of causal interaction is moot, "for *social properties are not immaterial or mental* any more than natural ones are" (Malm 2018, 65–66, emphasis in original).
5 According to Mike Dowie (2005), an estimated 20 to 50 million people have been evicted in the creation of protected areas at a global level (Dowie 2005). Earlier research (Geisler and De Sousa 2001), which focused exclusively on the case of Africa, put the number of "environmental refugees" at 15 million.

Bibliography

Armiero, Marco. 2015. "Ribelli. Naturalmente". In *La Contestazione Ecologica: Storia, Cronache e Narrazioni*, edited by Giorgio Nebbia (pp. 10–26). Naples: La scuola di Pitagora.

Avaaz. 2016. "Avaaz – World Leaders: Protect Half Our Planet". *Avaaz.* https://secure.avaaz.org/campaign/en/protect_half_our_planet_loc_sus/.

Barkham, Patrick. 2018. "'We're Doomed': Mayer Hillman on the Climate Reality No One Else Will Dare Mention". www.theguardian.com/environment/2018/apr/26/were-doomed-mayer-hillman-on-the-climate-reality-no-one-else-will-dare-mention.

Bonneuil, Christophe and Jean-Baptiste Fressoz. 2016. *The Shock of the Anthropocene: The Earth, History, and Us.* London: Verso.

Brush, Emma and Ted Nordhaus. 2018. "Nature Wars". *The Breakthrough Journal*, 9. https://thebreakthrough.org/journal/no.-9-summer-2018/from-the-editors2.

Büscher, Bram and Robert Fletcher. 2016. "Why E.O. Wilson Is Wrong about How to Save the Earth". Aeon. https://aeon.co/ideas/why-e-o-wilson-is-wrong-about-how-to-save-the-earth.

Chakrabarty, Dipesh. 2009. "The Climate of History: Four Theses". *Critical Inquiry* 35.

Dowie, Mark. 2005. "Conservation Refugees". *Orion Magazine*. https://orionmagazine.org/article/conservation-refugees/.

Dowie, Mark. 2009. *Conservation Refugees: The Hundred-year Conflict between Global Conservation and Native Peoples.* Cambridge, MA: MIT Press.

Geisler, Charles and Ragendra De Sousa. 2001. "From Refuge to Refugee: The African Case". *Public Administration and Development* 21(2): 159–170. doi:10.1002/pad.158.

Goldblatt, Colin and Andrew J. Watson. 2012. "The Runaway Greenhouse: Implications for Future Climate Change, Geoengineering and Planetary Atmospheres". January. doi:10.1098/rsta.2012.0004.

Hance, Jeremy. 2016. "Could We Set Aside Half the Earth for Nature?" *Guardian.* www.theguardian.com/environment/radical-conservation/2016/jun/15/could-we-set-aside-half-the-earth-for-nature.

Haraway, Donna. 2015. "Anthropocene, Capitalocene, Plantationocene, Chthulucene: Making Kin". *Environmental Humanities* 6(1): 159–165. doi:10.1215/22011919-3615934.

Hartley, Daniel. 2015. "Against the Anthropocene". *Salvage*. http://salvage.zone/in-print/against-the-anthropocene/#sdfootnote11sym.

Ingold, Tim. 1996. "Hunting and Gathering as Ways of Perceiving the Environment". In *Redefining Nature: Ecology, Culture, and Domestication*, edited by Roy Ellen and Katsuyoshi Fukui (pp. 117–155). https://philpapers.org/rec/INGHAG.

Ingold, Tim. 2000. "A Circumpolar Night's Dream". In *The Perception of the Environment: Essays on Livelihood, Dwelling and Skill* (pp. 89–110). London, New York: Routledge.

Kawa, Nicholas C. 2016. *Amazonia in the Anthropocene : People, Soils, Plants, Forests*. Austin: University of Texas Press. https://utpress.utexas.edu/books/kawa-amazonia-in-the-anthropocene.

Kingsnorth, Paul. 2013. "Dark Ecology". *Orion Magazine*. https://orionmagazine.org/article/dark-ecology/.

Latour, Bruno. 1993. *We Have Never Been Modern*. Cambridge, MA: Harvard University Press.

Latour, Bruno. 2010. "An Attempt at a Compositionist Manifesto". *New Literary History* 41: 471–490.

Latour, Bruno. 2014. "Agency at the Time of the Anthropocene". *New Literary History*. doi:10.1353/nlh.2014.0003.

Malm, Andreas. 2018. *The Progress of This Storm : Nature and Society in a Warming World*. London: Verso.

Marris, Emma. 2011. *Rambunctious Garden: Saving Nature in a Post-wild World*. London: Bloomsbury.

Mitchell, Timothy. 2002. *Rule of Experts: Egypt, Techno-politics, Modernity*. Berkeley, CA: University of California Press. doi:10.2307/23348139.

Monbiot, George. 2018. "We Won't Save the Earth with a Better Kind of Disposable Coffee Cup". *Guardian*. www.theguardian.com/commentisfree/2018/sep/06/save-earth-disposable-coffee-cup-green.

Moore, Jason W. 2016. *Anthropocene or Capitalocene. Nature, History and the Crisis of Capitalism. Anthropocene or Capitalocene. Nature, History and the Crisis of Capitalism*. Pm Press. doi:10.1017/CBO9781107415324.004.

Pollan, Michael. 1991/2003. *Second Nature. A Gardener's Education*. New York: Grove Press.

Pritchard, Sara B. 2012. "From Hydroimperialism to Hydrocapitalism: 'French' Hydraulics in France, North Africa, and Beyond". *Social Studies of Science* 42(4): 591–615. doi:10.1177/0306312712443018.

Scarso, Davide. 2013. "Beyond Nature and Culture?" *Limes: Borderland Studies* 6(2): 91–104. doi:10.3846/20297475.2012.753475.

Silberman, Neil A. 2013. "Heritage Interpretation as Public Discourse: Towards a New Paradigm". In *Understanding Heritage: Perspectives in Heritage Studies*, edited by Marie-Theres Albert, Roland Bernecker and Britta Rudolf. Berlin: De Gruyter. doi:10.1515/9783110308389.21.

Suckling, Kieran. 2014. "Against the Anthropocene". *Immanence: Ecolculture, Geophilosophy, Mediapolitics* (Blog). https://blog.uvm.edu/aivakhiv/2014/07/07/against-the-anthropocene/.

Tardieu, Christine. 2012. *Comment Nous Sommes Devenus Bipèdes: Le Mythe Des Enfants-Loups*. O. Jacob.

Wilson, Edward O. 2016a. *Half-Earth: Our Planet's Fight for Life*. New York: Liveright Publishing Corporation, a division of W.W. Norton.

Wilson, Edward O. 2016b. "The Global Solution to Extinction". *New York Times*. doi:10.1016/j.neuroimage.2008.02.053.

Wilson, Edward O. 2017. "A Biologist's Manifesto for Preserving Life on Earth". *Sierra Club Magazine*. www.sierraclub.org/sierra/2017-1-january-february/feature/biologists-manifesto-for-preserving-life-earth.

Wuerthner, George, Eileen Crist and Tom Butler, eds. 2014. *Protecting the Wild: Against the Domestication of Earth. Protecting the Wild: Parks and Wilderness the Foundation for Conservation*. Washington, DC: Island Press. doi:10.5822/978-1-61091-551-9.

Part III
Staging the Anthropocene

9 A new machine in the garden?

Staging technospheres in the Anthropocene

Nina Möllers, Luke Keogh and Helmuth Trischler

Introduction

Upon entering the first major exhibition on the Anthropocene which was jointly developed by the Rachel Carson Center for Environment and Society and the Deutsches Museum and shown in Munich from 2014 to 2016, the first thing that caught visitors' eyes was a large landscape complete with over 1,000 colourful paper flowers.[1] In the middle of this scenery stood a large cube made of metal beams holding almost 50 monitors. The aesthetic sensation created by this clash of colours, materials and forms was that of disruption and contrast.

In many ways, this architectural set-up of the opening segment referred back to the trope of the machine in the garden as described by Leo Marx in his seminal work of 1964 (Marx 1964). The disruption and permanent change of a seemingly pastoral landscape by the machinery of industrialization as featured in many prominent American nature writers, such as Henry David Thoreau, Herman Melville, Mark Twain and F. Scott Fitzgerald in the nineteenth and early twentieth century, has obviously gained new dimensions in the Anthropocene as a new geological epoch characterized by humans as the major driver in planetary change.

It is no coincidence that the curators and designers of the Munich Anthropocene exhibition chose to let visitors start their tour in such a dramatic architectural setting. And yet, it is not what it seems. Taking the well-known duality of nature and culture – in the shape of human-made technology – as a starting point, the exhibition explicitly goes beyond this binarism and invites its audience to experience the Anthropocene both as a geological debate and a philosophical and cultural concept that blurs the boundaries between well-established categories.[2]

In this chapter, we want to present the major conceptual ideas of the Munich Anthropocene exhibition by focusing on several selected elements of the gallery. These ideas showcase the underlying argument that the staging of the Anthropocene – both literally in an exhibition and figuratively in humans' actions of reshaping the Earth – resembles the act of designing and maintaining a garden. Although humans have catapulted themselves into a powerful driving position from which they are apparently able to change the Earth to their will,

Figure 9.1 View onto the flower landscape, media cube and Wall of Anthropocenic Objects in "Welcome to the Anthropocene. The Earth in Our Hands".

Photo: Courtesy of Deutsches Museum.

it is by no means a one-way street. Rather, we as humans are part of manifold interdependent structures of cause and consequence, so that the result of our behaviour and intentions can never be completely known to us. Moreover, the Anthropocene as a concept and discourse is not, as some critical voices are concerned about, gearing up for anthropocentrism (Malm and Hornborg 2014; Manemann 2014; LeCain 2015). Rather, by attributing agency to non-humans such as plants, animals and material objects, the concept fosters a de-centring of the *anthropos* in a post-humanist perspective (Latour 2014; Heise 2015). Although it may at first glance seem strange, this open-ended process of shaping the Earth is similar to that of garden design. Even in highly designed gardens, we know elements – both non-humans such as weeds, snails, greenflies and unruly humans – that destroy the order and idea of our garden. Similarly, the Anthropocene is shaped by a multitude of actors, often acting in concert, but just as often not. The Anthropocene as a garden is a matter of negotiation, contingency and unruliness, and as such it needs an open forum for discussion and decision-making.

The Wardian case: unwanted passengers in the garden

The Wardian case was a special wood and glass box used for transplanting plants around the globe. In 1829, the surgeon and amateur naturalist Nathanial

Figure 9.2 Wardian case, from the collection of the Botanic Garden and Botanic Museum Berlin-Dahlem, Freie Universität Berlin.

Photo: Courtesy of Deutsches Museum.

Bagshaw Ward accidentally discovered that plants enclosed in airtight glass cases can survive for long periods without watering. After the successful transport of plants to Australia and back the Wardian case was used for over a century to carry hundreds of thousands of plants around the globe. The Wardian case revolutionized the global movement of plants. Plants had greater chance of survival in the cases while in transit.

Certainly a product of the gardening crazes of the Victorian era, the Wardian case is also an artefact of the Anthropocene. The Wardian case resolved a major bottleneck in the transport of live plant species but also had major consequences for environmental relationships in the nineteenth and twentieth centuries. Invasion biologists working on the theme of biological globalization have reserved a special place for the case. Not only was the Wardian case a "milestone in botanical innovation", it was "a key vehicle for botanical globalization" (Weijden *et al.* 2007, 31–32).

In the Anthropocene our influence on mobility patterns has been widespread. Both the intentional and unintentional transport of species has had a major impact on the ecological histories of the planet. The humble shipping container that "made the world smaller" and moves "ninety percent of everything" appeared to curators as a sure symbol of globalization and the interconnections of the world system (Levinson 2007; George 2014). "The box" is an eloquent example of the Anthropocene and one we wanted to include in the Munich exhibition. But we also wanted to take people further back to some of the roots of our global mobility practices. The Wardian case was another box which was also a container and carrier of the Anthropocene, and that journey began in the garden.

Whether carrying prized horticultural exotics or useful economic plants, the case was an important prime mover. Some of the key uses of the case include moving tea from China to India to lay the foundations of the Assam and Darjeeling tea districts; stealing rubber from Brazil and transporting it via London to Asia which is now the leading producer of the crop; and repeatedly moving bananas over many decades to the Pacific Islands, Central America and the Caribbean.

Bananas offer a case in point of the dramatic and complex world of moving species. In much of the early nineteenth century the Gros Michel variety of banana was the dominant variety of banana, but after suffering from Panama disease the Cavendish variety became the leading type of banana. Plantation bananas are all grown from clones. Today, the ecologist Rob Dunn (2017) argues that the Cavendish banana as a collective organism may well be the largest organism on Earth. Its global rise came about from our ability to move plants efficiently and effectively. But there have always been costs. Today, the Cavendish banana is under threat of disease and could soon be wiped out.

The shift towards monoculture plantations to supply the global economy – many of these like rubber and cacao set up with the use of the Wardian case – created greater susceptibility to disease and invasion. Plant diseases were widely circulated in Wardian cases. The sugarcane mosaic virus and coffee rust are two examples that were found in cases and devastated crops. People did not reject monocultures but instead looked to quarantine and controlling the movement of species. And while these may be effective to some degree we are still suffering the consequences of the lag effect of invasive species transport and transformation.

Non-native species, including many plants, are now one of the major drivers of global biodiversity loss and economic impacts for the globe are as high as US$120 billion annually (Keller *et al.* 2011; Pimental *et al.* 2005). Due to human activities, of all the known vascular plants on Earth, at least 3.9 per cent have become naturalized outside their natural ranges (van Kluenen *et al.* 2015). Humans, more than natural forces, are now the largest dispersers of vascular plants in the global garden (Mack and Lonsdale 2001).

Myriad machines in the garden: the creation of a technosphere

The geologist Peter K. Haff has used the term "technosphere" to allow for a new look at the role of humans and human purpose in the Anthropocene. According to him, the term "can help avoid the misleading anthropocentric assumption that humans are independent authors of the Anthropocene they inhabit, rather than contingent actors whose purposes are not entirely their own" (Haff 2016, 54). The hybrid human-technological system of the technosphere consists of:

> [T]he world's large-scale energy and resource extraction systems, power generation and transmission systems, communication, transportation, financial and other networks, governments and bureaucracies, cities, factories, farms and myriad other "built" systems, as well as all the parts of these systems, including computers, windows, tractors, office memos.

It also includes social and cultural systems such as religious institutions or non-governmental organizations (Haff 2014, 127). By conceptualizing the technosphere as an emergent system with its own internal dynamics that establishes itself increasingly beyond human control, Haff builds on the trope of an autonomous technology that philosophers and sociologists of technology such as Jacques Ellul and Langdon Winner developed almost half a century ago (Ellul 1967; Winner 1977). In fact, according to the *Oxford English Dictionary*, the term "technosphere" dates back well into the 1960s.

While Haff understands the technosphere in a temporal sense, basically synonymous with the Anthropocene as a new epoch in geological times, most scholars prefer to use it in a spatial sense, as the term "sphere" implies.[3] The interdisciplinary body of the Working Group on the Anthropocene has even gone so far as to quantify the technosphere. In a recent publication, the group estimates the technosphere's mass at approximately 30 trillion tonnes, no less than around five orders of magnitude bigger than the ever-growing human biomass (Zalasiewicz *et al.* 2016).

No doubt this gigantic technical machinery in the garden named Earth is as massive as it is new. It has transformed much of the Earth's surface into a technosphere of global scale and "its myriad components underscore the novelty of the current planetary transformation", as members of the Working Group on the Anthropocene stress (Zalasiewicz *et al.* 2016, 9).

In search of a beginning

When did this gigantic sphere of human-built technologies emerge? The answer to this question is complex. It weaves together into a rich fabric elements of *long durée* processes in human history that date back well into the beginnings of the Holocene with recent phenomena in technology-based societies. The colourful tapestry began to take shape in the so-called Neolithic Revolution with the transition from nomadic societies of hunters, gatherers and fishers to permanent settlements engaged in agriculture some 11,700 years ago. The "invention" of sedentary societies, agriculture, and animal and plant husbandry was closely connected with new technologies: firing pottery made it possible to store agricultural produce for later use. Improved stone tools and innovations in building methods provided a basis for permanent settlements. The new machine that probably had the biggest impact on the garden Earth was the plough that allowed humans to break up the soil for cultivation and increased productivity. The serial character of technical innovations distinguishes the Neolithic period fundamentally from earlier periods in the development of humankind. The technosphere began to take shape.

The Neolithic Revolution took 5,000 years to unfold its transformative power beyond its spatial origins in the Fertile Crescent in the Near East, a true Garden of Eden in early antiquity. Over time, then, the momentum shifted from East to West, and Western societies developed what historian of technology Robert Friedel (2007) has called a "culture of improvement". The evolutionary process of incremental innovations gradually enlarged the technosphere in the course of the second millennium. It culminated in the Industrial Revolution that started in Great Britain in the mid-eighteenth century. Technological innovation was the driving force in this revolution. Three main processes worked together: first, the mechanization of manual labour with textile production at the beginning; second, mechanical production and transformation of energy using the steam-engine; and third, the large-scale exploitation, production and use of coal and iron. The first phase of the Industrial Revolution had already created an economy and society built on scientific and technical knowledge, even before the second wave of industrialization towards the end of the nineteenth century, when chemistry and electrical engineering brought with them a breakthrough into modern innovation systems based on scientific and technical expertise (North 1990; Mokyr 2002).

The *Welcome to the Anthropocene* exhibition visualized the dynamic growth of the technosphere during industrialization in a "Wall of Anthropocenic Objects". The 20-metre-long display was made of a paper product. The honeycomb structure both enclosed and revealed iconic objects such as a steam-engine, a tractor, an aircraft engine, a satellite, and a combination television and media chest that represent technical milestones on the road to the Anthropocene. The wall's surface was artistically covered in handwritten notes and drawings that pointed out the many links and intersections among technologies. Neither these intersections nor the resulting transformative dynamics were fully foreseeable to the contemporaries of industrialization. The technosphere consists of human-made artefacts, but humanity has learned that this emergent and dynamically growing sphere often grows beyond control. Technical disasters in the nuclear energy business such as Harrisburg, Chernobyl and Fukushima are the most visible examples. The lightweight, paper-based and – surely, only seemingly – fragile structure of the wall, with its handwritten labels that allowed for easy changes, aimed to represent both the unruliness of the new machines in the garden and the openness of the Anthropocene debate.

Large-scale machines such as steam-engines and tractors are potential trace fossils of the Anthropocene. In a distant future, when humans as a biological species will most probably already have become extinct, they will end up in a stratigraphic layer of techno-fossils that marks the epoch of the Anthropocene. Today, however, the technosphere consists not only of bulky physical objects

Figure 9.3 Detail of the Wall of the Anthropocenic Objects showing a Lanz Bulldog HL 12, manufactured by Heinrich Lanz, Mannheim in 1921.

Photo: Courtesy of Deutsches Museum.

but also of equally suitable objects such as genetically modified tomatoes or cloned sheep. The Earth is filled with such hybrid sociotechnical objects/species that undermine the traditional distinction of living organisms and inanimate artefacts and extend the technosphere into a bio-geo-technosphere (Möllers 2015b). Moreover, geologists have identified invisible objects as potential stratigraphic signatures of the Anthropocene. The concentration of greenhouse gases in the atmosphere is a particular strong candidate to signify the "age of humans"; another is the large quantity of radionuclides released by nuclear bomb testing. The fallout from the latter, some suggest, marked the prehistory of the Anthropocene, commencing in the late 1930s, when Otto Hahn, Lise Meitner and Fritz Straßmann discovered nuclear fission while looking for transuranium elements at the Kaiser-Wilhelm-Institute for Chemistry in Berlin. The experimental set-up of the Berlin team later found its way into the collections of the Deutsches Museum and went on display at the *Welcome to the Anthropocene* exhibition. These objects helped showcase the early traces of humanity as a geological factor: the altering of the face of the Earth and its atmosphere through the use of nuclear technologies.

In their search for suitable markers of the new geological epoch under consideration, the Working Group on the Anthropocene has also turned to the nuclear sector. Radionuclides resulting from atomic tests would allow the precise date of the beginning of the Anthropocene as 16 July 1945, when the first deto-nation of a nuclear explosive, the "Trinity Test", took place on a US Army missile range near Alamogordo in the New Mexico desert. Environmental historian and fellow Working Group member John McNeill, however, recom-mended to geologists in need of a clear-cut golden spike the examination of the bones and teeth of mammals born in the 1940s and 1950s, which for the first time in history include a chemical signature that resulted from nuclear weapons tests (McNeill and Engelke 2016). Some of these bones and teeth will end up in a sediment layer that distinguishes the mid-twentieth century from all that went before and all that came after, because the nuclear test ban treaty of 1963 then weakened the signature of radionuclides.

The nature of the future

The nature of future fossils resulting from the epoch of the Anthropocene is not only of interest to geologists. The concept of the Anthropocene has also proved attractive to artists, who are exploring the implications of deep time changes on how people respond to our changing environment. The curatorial team of the *Welcome to the Anthropocene* gallery decided to collaborate in particular with people who aim to create a new hybridity between science, technology and the arts and to incorporate in themselves an expertise both in the arts and sciences. The works of two of these hybrid scholars, who in their artistic creations specu-late about the future, are worthy of further discussion.

What is the relationship between the biosphere and the growing technosphere shaped by humans? Is the development of the five-bladed razor the outcome of human needs? The Netherlands-based art and design group Next Nature is deter-mined to explore such questions. They created *Razorius gillettus* – a new species in our techno-economic ecology. The development of razor technology resembles natural evolutionary processes: each model builds upon the properties of previous models. Successful adaptations are preserved in future generations, while unsuc-cessful ones disappear. Aesthetic properties often have no obvious function except that they help one model compete against its rivals. One could argue that razors are man-made objects and they are unable to reproduce themselves. But this is also true today in the case of many domesticated fruits and livestock.

Another of Next Nature's projects artistically repurposes space blankets for use during natural disasters. As the severity of extreme events increases due to human activity, they pose the question of whether we will be able to survive nature. Originally developed by the US space agency NASA in 1964 to protect astronauts in outer space, the tear-resistant and waterproof blanket is made of polyester coated with a thin layer of reflective aluminium. Both in space and on Earth, it protects its user from the forces of "old" nature: cold, heat, rain and wind. But this special emergency blanket is different. It is meant to protect us from the forces of "new" natures in the ever-growing technosphere: electrosmog, drone attacks or radio waves. It asks us to consider: Are we at odds with nature – or with ourselves?

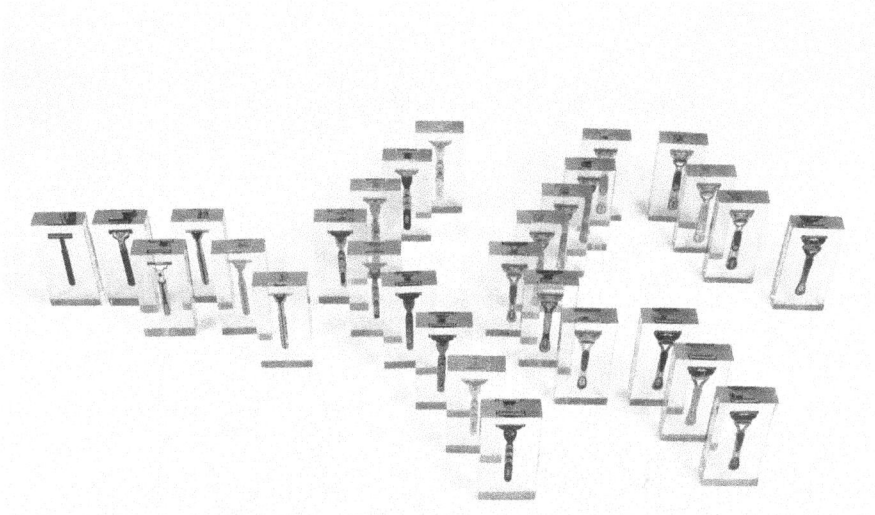

Figure 9.4 Razorius Gilletus (2014) by Next Nature Network.
Photo: Courtesy of Deutsches Museum.

For the artist-engineer-designer scholars of Next Nature there is no doubt that the human-created technosphere is a technological wilderness implying economic and ecological crisis. The technosphere, however, is not necessarily bound to become a lifeless wasteland. Rather, the works of Next Nature envision a practical utopia "where humans function not as bumbling builders of risky machine, but as the benevolent catalysts of evolution" (van Mensvoort and Guy 2015, 50).

The London-based artist and designer Yesenia Thibault-Picazo investigates possible man-made geological formations of the future and in doing so collaborates closely with the geologist Jan Zalasiewicz, head of the Working Group on the Anthropocene, to secure scientific plausibility of her artwork. Her ingenious "Anthropocenic Specimen Cabinet", a material library of future geology, starts from the observation that humanity has spread specific elements in nature which were rare in the pre-human era, but which will become prevalent sediments in future planetary strata. She elaborates "material tales that speculate about substances that might be mined in a far future" (Thibault-Picazo 2015, 114). Such future fossils allow visitors to contextualize the premises of the Anthropocene concept and to project the long-term temporal impact of current activities. The pieces of her display, produced in a laboratory-like set-up using the technique of geo-mimicry, serve as time capsules from an imagined future. They link the deep history of sedimentation processes in Earth history with the long present of the durability of human-produced waste such as plastics and the deep future of transforming human-altered substances into techno-fossils.

Figure 9.5 Survival blanket (2014) by Next Nature Network.
Photographer: Axel Griesch. Photo courtesy of Deutsches Museum.

Gardeners of time

Temporalities are at stake in the overall debate about the Anthropocene. Much intellectual energy has been spent in reflecting on possible temporal starting points of the Anthropocene and the geological markers that would proof the new epoch. For the stratigraphy community this is the primary task of the Working Group on the Anthropocene: to identify a synchronous base which is the same time everywhere around the globe and accordingly a specified position in the sedimentary record that defines this synchronous base, a so-called Global Boundary Stratotype Section and Point (GSSP), known as a "golden spike". The Working Group has been entrusted with the task of examining the scientific evidence for each of these suggestions and submitting a formal proposal based on their own stratigraphic investigations to the Subcommission on Quaternary Stratigraphy. This organization, in turn, reports to the International Commission on Stratigraphy, which reports to the International Union of Geological Sciences. Only after the proposal has successfully passed through all four of these scientific bodies with a qualified majority will the existence of a new geological division be considered official.

Meanwhile the Working Group has come to the conclusion that "the Anthropocene is functionally and stratigraphically distinct from the Holocene". The group found stratigraphic signatures that were either completely new, or fell substantially outside the level of variation of the Holocene; all of these changes were occurring in an accelerated fashion (Waters *et al.* 2016). The data collected suggested that the lower boundary of the Anthropocene should be drawn in the mid-twentieth century, which was named earlier as the "Great Acceleration".

This temporal proposal is, however, far from having found a consensus across academic disciplines and research fields. Historians stress that the idea of an epoch marked by far-reaching human influence on the Earth is much older. They have tracked precursors of the idea of humans as geological agents back to the late eighteenth century, when, in 1775, for example, the French naturalist Georges-Louis Leclerc, Comte de Buffon, distinguished between original nature and nature civilized by humans, and observed that the entire face of the Earth bears the imprint of human power (Leclerc 1778).[4]

Stratigraphers and historians are both true gardeners of time. Both scholarly communities seek to take care of the garden of temporalities, albeit directed towards different temporal scales and equipped with different tools to conduct the business of periodization. While the former community focuses on big history usually measured in hundreds and thousands if not millions of years, historians aim at getting a better understanding of recent pasts. The major challenge of the Anthropocene concept is to bring together the seemingly incommensurable temporal scales of geological times and human times. Calling for a convergence of natural and human temporalities, the Anthropocene debate asks its contributors to remain open to venturing out to new temporal concepts and epistemologies of historical change. The debate also invites them to experiment with innovative tools to represent change in order to question established trails and pathways in the garden of time.

A fascinating experiment to visualize the merger of geological and human times provides the "Clock of the Long Now". In order to recognize the potential of the Anthropocene to open up new perspectives by juxtaposing vastly different

Figure 9.6 "Clock of the Long Now" installation.
Photo: Courtesy of Deutsches Museum.

time scales, scholars do not necessarily have to look as far ahead as to the end of the Earth in several billion years, or backwards to the Big Bang 13.7 billion years ago that began everything, as proponents of "Big History" advocate (Christian 2005). Rather, the Anthropocene combines the long history of human alteration of the natural environment starting during the Neolithic Revolution with the long present, made tangible in the form of the "Clock of the Long Now". This ongoing experiment to visualize the merger of geological and human times was initiated by California cyber-utopian Steward Brand and his Long Now Foundation. The aim is to build a clock that will keep time for 10,000 years or more without the need for human maintenance. The model of this apparatus, displayed in *Welcome to the Anthropocene*, represents the dialectical constellation that humans must take responsibility for the consequences of their actions, which will continue into the unimaginably distant future, without knowing how they can adequately fulfil this responsibility (Brand 1999; Möllers 2015a, 114).

This dialectic is also evident, for example, in the law passed by the German government regarding the selection of a long-term repository site for radioactive waste. The law requires that such a storage site must ensure safety for one million years. This desperate attempt borders on hubris, reminding us that high-level radioactive waste is a substance that we cannot control, but rather, as legal scholar Jens Kersten notes, "will control us – at least insofar as we continue to exist in the coming million years" (Kersten 2016, 285, 2013).

Flower garden of Anthropocenic futures

A garden gives the feeling that it is controlled by humans, but we just tend it; it is filled with many species other than human, all having an impact. It is mostly neat and tidy, as, for example, rows of maize in the US Mid-West, palm oil plantations in Indonesia or rice paddies in China. Boreal woods in northern Finland, deserts in Africa or tropical rainforests in Amazonia are less garden-like. Even these remote places are, however, also impacted by human agency. Fast-growing CO_2 emissions and slowly rising temperatures have influenced these ecosystems, forcing them to adjust to human-caused climate change. Humans have neither planned nor organized the transformation of these places – it has happened contingently, slowly and in an unruly fashion. This is why the writer, Anthropocene scholar and passionate gardener Emma Marris (2011) staged the human-altered Earth as a "rambunctious garden".

The metaphor of the Earth as a rambunctious garden seeks to capture the ambivalent nature of the Anthropocene. It points to the transformative power of humanity while emphasizing continuing elements of wild natures and uncontrolled technologies. As with most metaphors, the semantic figure of the rambunctious garden can only partly grasp the complexity of the nature–culture relations during the period of an ever-expanding technosphere, the primary reason of humanity forcefully developing into a geo-biological factor. Surely, Marris is aware that the Earth as a garden cannot be controlled "like an engineer running a machine" (Marris 2015, 45). The ongoing controversy about the Good Anthropocene initiated by Erle C. Ellis (2011) underlines that the Anthropocene is a conceptual space where fundamental ethical positions are constantly being renegotiated. Ellis, a US geographer and landscape ecologist, recognizes that "the only limits to creating a planet that future generations will be proud of are our imaginations and our social systems". He continues: "In moving toward a better Anthropocene, the environment will be what we make it" (Ellis 2015, 54).

The idea of the Good Anthropocene has incurred wide criticism from those who fear that it paves the way for questionable concepts of geoengineering (Latour 2012; Hamilton 2015). Critics point out that several years ago, no less a figure than the 'father' of the Anthropocene, Paul J. Crutzen, despairing about the effects of anthropogenic climate change, published the suggestion of stopping global warming by injecting 1.5 million tonnes of sulphur dioxide particles into the atmosphere in order to reflect sunlight (Crutzen 2006). Crutzen has been heavily criticized for this. As a consequence of such talk of technological interventions when the long-term effects on the climate and the Earth are completely unknown, the idea of the Anthropocene has acquired for many a bitter aftertaste (Hamilton 2013).

The controversial debate about the Good Anthropocene manifests that at stake is nothing less than the most central questions of our society: What role will technology have in this? What forms of production and communication of knowledge are suitable for the Anthropocene? How should we conduct business, work and live? After all, what will the Anthropocenic Earth look like in 100 years from now?

These were the questions with which the *Welcome to the Anthropocene* exhibition ended. The curators invited key experts and stakeholders of the Anthropocene debate such as the climate scientist Will Steffen, the environmental historian Christof Mauch, the geologist Jan Zalasiewicz, the young environmental activist and founder of "Plant for the Planet" Felix Finkbeiner, and the artist and designer Koert van Mensvoort to answer these questions and provide their statements in a media station that visitors could listen to.

Despite the importance of experts steering us all through an increasingly complex world, the message with which our exhibition ended was that of individual relevance and personal feeling. Aside from systemic dependencies, an intricate interplay between micro and macro level characterizes the manifold phenomena of the Anthropocene bio-geo-technosphere. Both the global collective and the local individual shape the new epoch of planet Earth, which is why one of the prime messages of our exhibition was: You are the Anthropocene! No matter how small we may sometimes feel, the Anthropocene makes each and every one of us global players. Even in the moment where we decide to do nothing, we are part of the Anthropocene. Personal and emotional reactions to issues and problems are basic prerequisites for turning awareness into responsible action. The encouragingly high motivation of visitors to stay informed and become active in shaping the Anthropocene was demonstrated

Figure 9.7 A visitor adds his thoughts on the Anthropocene to the exhibition's flower garden.

Photographer: Axel Griesch. Photo courtesy of Deutsches Museum.

both in the findings of a scientific evaluation of the exhibition and in thousands of thoughtful, balanced, funny, provocative and always inspiring comments that our visitors left in the gallery's flower garden. Here, we invited them to become symbolic gardeners of the Anthropocene by writing down their thoughts, fears, hopes, wishes and ideas on a coloured piece of paper and adding it – folded like a flower – to the collective garden. Over the duration of the exhibition, it was periodically harvested by the curators. The tens of thousands of responses collected were bound into books and also displayed in the exhibition, and they form an invaluable stockpile of social feedback to the idea and phenomena of the Anthropocene. Concern and fear, sometimes anger, resonate with many comments, but also a great amount of hope and spirit to tackle the manifold problems we may face in an age where humans have become major forces of nature. Perhaps a little surprisingly, many younger visitors in particular chose to spend some time with this participatory element, often writing page-long commentaries. The visitors' comments ranged from grand dreams and visions for the future to small but indispensable acts of change which they promised to tackle in their daily routines. Most agreed on the assessment that it is high time to take up the responsibilities that we ourselves have created and that it will be demanding. Summing it up rather perfectly, one visitor – perhaps an avid gardener him- or herself – wrote laconically: "The Anthropocene creates a lot of work. Oh well, anyway!"

Figure 9.8 Example of a visitor's flower in the Anthropocene exhibition.

Photo: Courtesy of Deutsches Museum.

Notes

1 *Welcome to the Anthropocene. The Earth in Our Hands* (German title: Willkommen im Anthropozän. Unsere Verantwortung für die Zukunft der Erde), Deutsches Museum, München, December 2014 to September 2016. An in-depth view of the exhibition, its conceptual design and its creation is offered by Möllers *et al.* (2015); see also Robin *et al.* (2014, 2017).

2 For a discussion of the Anthropocene as a "trading zone" of sciences and humanities see Trischler (2016).

3 For an overview on recent research see Trischler and Will (2017) and the *Techno-sphere Magazine*, edited by the Haus der Kulturen der Welt in Berlin: http:// technosphere-magazine.hkw.de/#/?_k=lwy995 (accessed 4 April 2017). For the idea of the "Technocene" see Hornborg (2015).

4 An analysis of the term's conceptual origins is provided by Schwägerl (2015), while the idea of an intellectual prehistory of the term is furiously criticized by Hamilton and Grinevald (2015).

Bibliography

Brand, Steward. 1999. *The Clock of the Long Now: Time and Responsibility*. New York: Weidenfeld & Nicolson.

Christian, David C. 2005. *Maps of Time: An Introduction to Big History*. Berkeley: University of California Press.

Crutzen, Paul J. 2006. "Albedo Enhancement by Stratospheric Sulfur Injections: A Contribution to Resolve a Policy Dilemma?" *Climatic Change* 77: 211–220.

Dunn, Rob. 2017. *Never Out of Season: How Having the Food We Want When We Want it Threatens Our Food Supply and Our Future*. New York: Little, Brown.

Ellis, Erle. 2011. "The Planet of No Return". *Breakthrough Journal*, autumn. Available at http://thebreakthrough.org/index.php/journal/past-issues/issue-2/the-planet-of-no-return (accessed 28 March 2017).

Ellis, Erle. 2015. "The Used Earth: Embracing Our History as Transformers". In *Welcome to the Anthropocene: The Earth in Our Hands*, edited by Nina Möllers, Christian Schwägerl and Helmuth Trischler (pp. 52–55). Munich: Deutsches Museum.

Ellul, Jacques. 1967. *The Technological Society*. New York: Vintage.

Friedel, Robert. 2007. *A Culture of Improvement: Technology and Western Millennium*. Cambridge, MA, and London: The MIT Press.

George, Rose. 2014. *Ninety Percent of Everything: Inside Shipping, the Invisible Industry that Puts Clothes on Your Back, Gas in Your Car, and Food on Your Plate*. London: Picador.

Haff, Peter K. 2014. "Humans and Technology in the Anthropocene: Six Rules". *The Anthropocene Review* 1(2): 126–136.

Haff, Peter K. 2016. "Purpose in the Anthropocene: Dynamical Role and Physical Basis". *Anthropocene* 16: 54–60.

Hamilton, Clive. 2013. *Earthmasters: The Dawn of the Age of Climate Engineering*. New Haven, CT: Yale University Press.

Hamilton, Clive. 2015. "The Theodicy of the 'Good Anthropocene'". *Environmental Humanities* 7: 233–238.

Hamilton, Clive and Jan Grinevald. 2015. "Was the Anthropocene Anticipated?" *The Anthropocene Review* 2(1): 59–72.

Heise, Ursula. 2015. "Posthumanism: Reimagining the Human". In *Welcome to the Anthropocene: The Earth in Our Hands*, edited by Nina Möllers, Christian Schwägerl and Helmuth Trischler (pp. 38–42). Munich: Deutsches Museum.

Hornborg, Alf. 2015. "The Political Ecology of the Technocene: Uncovering Ecologically Unequal Exchange in the World-system". In *The Anthropocene and the Global Environmental Crisis. Rethinking Modernity in a New Epoch*, edited by Clive Hamilton *et al.* (pp. 57–69). New York: Routledge.

Keller, Reuben P., Jürgen Geist, Jonathan M. Jeschke and Ingolf Kuhn. 2011. "Invasive Species in Europe: Ecology, Status, and Policy". *Environmental Sciences Europe* 23: 1–17.

Kersten, Jens. 2013. "The Enjoyment of Complexity: A New Political Anthropology for the Anthropocene?" In *Anthropocene: Exploring the Future of the Age of Humans* (RCC Perspectives 2013/3), edited by Helmuth Trischler (pp. 39–55). Munich: Rachel Carson Center.

Kersten, Jens. 2016. "Eine Million Jahre? Über die juristische Metaphysik der atomaren Endlagerung". In *Inwastement: Abfall in Umwelt und Gesellschaft*, edited by Jens Kersten (pp. 269–287). Bielefeld: transcript.

Kluenen, Mark van, Wayne Dawson, Franz Essl, *et al.* 2015. "Global Exchange and Accumulation of Non-native Plants". *Nature* 525: 100–103.

Latour, Bruno. 2012. "Love Your Monsters: Why We Must Care for Our Technologies as We Do Our Children". *Breakthrough Journal* winter. Available at http://the breakthrough.org/index.php/journal/past-issues/issue-2/love-your-monsters (accessed 28 March 2017).

Latour, Bruno. 2014. "Agency at the Time of the Anthropocene". *New Literary History* 45: 1–18.

LeCain, Timothy J. 2015. "Against the Anthropocene: A Neo-materialist Perspective". *International Journal of History, Culture, and Modernity* 3: 1–28.

Leclerc, Georges-Louis. 1778. *Histoire Naturelle, générale et particulière, avec la description du Cabinet du Roi, Vol. 34 (Supplement 5)*: Des Époques de la nature. Paris: Imprimerie Royale.

Levinson, Mark. 2007. *The Box: How the Shipping Container Made the World Smaller and the World Economy Bigger*. Princeton, NJ: Princeton University Press.

Mack, Richard N. and Mark W. Lonsdale. 2001. "Humans as Global Plant Dispersers: Getting More Than We Bargained For". *BioScience* 51 (2): 95–102.

Malm, Andreas and Alf Hornborg. 2014. "The Geology of Mankind? A Critique of the Anthropocene Narrative". *The Anthropocene Review* 1(1): 62–69.

Manemann, Jürgen. 2014. *Kritik des Anthropozäns. Plädoyer für eine neue Humanökologie*. Bielefeld: transcript.

Marris, Emma. 2011. *Rambunctious Garden*. New York: Bloomsbury.

Marris, Emma. 2015. "The Global Garden". In *Welcome to the Anthropocene: The Earth in Our Hands*, edited by Nina Möllers, Christian Schwägerl and Helmuth Trischler (pp. 43–46). Munich: Deutsches Museum.

Marx, Leo. 1964. *The Machine in the Garden. Technology and the Pastoral Ideal in America*. Oxford: Oxford University Press.

McNeill, John and Peter Engelke. 2016. *The Great Acceleration. An Environmental History of the Anthropocene since 1945*. Cambridge, MA: Harvard University Press.

Mensvoort, Koert van and Allison Guy. 2015. "The Anthropocene Explosion". In *Welcome to the Anthropocene: The Earth in Our Hands*, edited by Nina Möllers, Christian Schwägerl and Helmuth Trischler (pp. 47–50). Munich: Deutsches Museum.

Mokyr, Joel. 2002. *The Gifts of Athena. Historical Origins of the Knowledge Economy*. Princeton, NJ: Princeton University Press.

Möllers, Nina. 2015a. "The Clock of the Long Now". In *Welcome to the Anthropocene: The Earth in Our Hands*, edited by Nina Möllers, Christian Schwägerl and Helmuth Trischler (pp. 114–119). Munich: Deutsches Museum.

Möllers, Nina. 2015b. "The Bio-Geo-Technosphere: Humans and Machines in the Anthropocene". In *Welcome to the Anthropocene: The Earth in Our Hands*, edited by Nina Möllers, Christian Schwägerl and Helmuth Trischler (pp. 154–159). Munich: Deutsches Museum.

Möllers, Nina, Christian Schwägerl and Helmuth Trischler (eds). 2015. *Welcome to the Anthropocene: The Earth in Our Hands*. Munich: Deutsches Museum.

North, Douglass C. 1990. *Institutions, Institutional Change, and Economic Performance*. Cambridge: Cambridge University Press.

Pimental, David, Rodolfo Zuniga and Doug Morrison. 2005. "Update on the Environmental and Economic Costs Associated with Alien-invasive Species in the United States". *Ecological Economics* 52(3): 273–288.

Robin, Libby, Dag Avango, Luke Keogh, Nina Möllers and Helmuth Trischler. 2017. "Displaying the Anthropocene In and beyond Museums". In *Curating the Future. Museums, Communities and Climate Change*, edited by Jennifer Newell, Libby Robin and Kirsten Wehner (pp. 252–266). London and New York: Routledge.

Robin, Libby, Dag Avango, Luke Keogh, Nina Möllers, Bernd Scherer and Helmuth Trischler. 2014. "Three Galleries of the Anthropocene". *The Anthropocene Review* 1(3): 207–224.

Schwägerl, Christian. 2015. "A Concept with a Past". In *Welcome to the Anthropocene: The Earth in Our Hands*, edited by Nina Möllers, Christian Schwägerl and Helmuth Trischler (pp. 28–30). Munich: Deutsches Museum.

Thibault-Picazo, Yesenia. 2015. "Crafts in the Anthropocene. Fossils from the Future". In *Welcome to the Anthropocene: The Earth in Our Hands*, edited by Nina Möllers, Christian Schwägerl and Helmuth Trischler (pp. 114–119). Munich: Deutsches Museum.

Trischler, Helmuth. 2016. "The Anthropocene – A Challenge for the History of Science, Technology, and the Environment". *N.T.M. – Journal of the History of Science, Technology, and Medicine* 24(3): 309–335.

Trischler, Helmuth and Fabienne Will. 2017. "Technosphere, Technocene, and the History of Technology". *ICON: Journal of the International Committee for the History of Technology* 23: 1–17.

Waters, Colin N., Jan Zalasiewicz, Colin Summerhayes, *et al.* 2016. "The Anthropocene is Functionally and Stratigraphically Distinct from the Holocene". *Science* 351(6269).

Weijden, Wouter van der, R.J. Lewis and Pieter Bol. 2007. *Biological Globalisation: Bioinvasions and Their Impacts on Nature, the Economy and Public Health*. Amsterdam: KNNV.

Winner, Langdon. 1977. *Autonomous Technology: Technics-out-of-control as a Theme in Political Thought*. Cambridge, MA: MIT Press.

Zalasiewicz, Jan, Mark Williams, Colin N. Waters, *et al.* 2016. "Scale and Diversity of the Physical Technosphere: A Geological Perspective". *The Anthropocene Review* 3(2): 9–22.

10 The atom in the garden and the apocalyptic fungi

A tale on a global nuclearscape (with artworks and bird-songs)

Jaume Valentines-Álvarez with Eric LoPresti

To little-clover Rita, Max and Noa at dawn
To grandtree Balbina at dusk
To flower Flor at night

Preface

This "tale" – illustrated with pictures, artworks, songs and songbirds – tries to reflect upon the history of the making of a nuclear landscape on a global scale throughout the twentieth and into the beginning of the twenty-first century. It is a personal and political account about how nuclear technologies shaped and still shape our real, imaginary, utopian and dystopian landscapes. It is indebted to anarchist thought, texts and practices, and has also benefited from lived experiences, art perspectives and different academic traditions such as the history of technology, philosophy, anthropology, geology and microbiology.

The "tale" brings different textual, visual and acoustic forms of expression together in order to explore new historiographical ways of storytelling. Due to publishing constraints, images have been restricted to two colourful paintings of the series "Blooms" by Eric LoPresti, which were exhibited at the Elizabeth Houston Gallery in New York City in 2017. Besides being accompanied by visual resources, the text has been written to be read (or to be heard) while the music, audiovisuals or sounds recommended below the sections' titles are playing (the recording times are approximately the times needed to read every section; all the tracks can be easily found online).

The sections conducted by music songs and film songs ("#Play Track") introduce – more or less chronologically, less or more in a simplified way – different moments of the story: Section 1, "The Forbidden Fruit in a Globescape" (around 1929–1945), Section 2, "The Gospel of Nuclear Edens" (1953–1986), Section 3, "Utopians are Everywhere" (1973–1986), and Section 4, "Updating: Nuclear_ Eden.02" (2000 to the present) (as observed, the years around the biggest global financial and energetic crisis – 1929, 1973 and 2008 – play an important role). The two shorter sections conducted through sounds ("#Play Sounds"), are devoted to art, and especially to the works by LoPresti in New York and the West Coast collective in Lisbon.

The music selection has been made with a view to the need to complement and contextualize the text (like in the artworks presented, melodies and lyrics contain epistemic clues for this story). The Opening and the Finale sections of the "tale" – which are devoted to thinking about what fungi teach us about life renewal – refer to the Greek myth of the female oracles called sibyls. This myth was appropriated during the Christian medieval world to prophesy the last judgment and the end of the world. The first track is a medieval Catalan version of the Song of the Sibyl recorded by soprano Montserrat Figueras and Jordi Savall. Figueras reviewed its meaning in the following terms (Figueras and Savall 1996):

> The Sibyl's devastating words are dramatically relevant today, speaking as they do of the destruction of the planet, of mankind's lack of respect for the vanishing natural world and of the brutality that has led man to regard nature as a machine.

The last track is based on the version of the Song of the Sibyl which is still sung in Majorca on Christmas Eve. It was recorded by the Iberian percussion orchestra Coetus, which means – in Latin – crowd, political or illegal assembly, or union. At the end of this recording, whistles and musical instruments reproduce bird sounds as a metaphor of the possibility of new worlds.

"We are not in the least afraid of ruins. We are going to inherit the Earth. [...] We carry a new world, here in our hearts. That world is growing this minute" are the famous words by anarchist Buenaventura Durruti (León, 1896-Madrid, 1936). The new worlds that "we" carry in our hearts are cooperative refuges for life based on biological and social mutual aid, instead of shelters based on technocratic-based technologies, forced security and authority. Rosalind Williams pointed out in the final lines of *Notes on the Underground* (Williams 2008, 213):

> The human environment is by definition technological to some degree. But if we allow technology to take over our surroundings, they can become inhospitable to human life. [...] Our increased dependence on technological shelter may lead to the weakening of human interdependence, which is another source of security. We should not forget that society too provides shelter, and in many cases a more flexible and effective kind.

Opening. The Apocalyptic Fungus (I The Mushroom)[1]

#PLAY TRACK. Monserrat Figueras, "El cant de la Sibil·la" [Barcelona Sibyl] (10/11th centuries) [18:13]#

A single large mushroom such as a specimen of *Langermannia gigantea* or of *Macrolepiota procera* (popularly known as Giant puffball and Parasol) can bear several billions of tiny spores on its fruiting body and can spread tens of thousands of them every second. Air currents can make these spores blow to a height up to tens of miles into the atmosphere during days, weeks and months, and cross mountains, cities, seas, and even oceans. If all the spores of a single large mushroom of these species found a proper milieu, they could colonize the whole of the Earth's surface.

In 1945, the several-thousand-metre-high mushroom of gases, dust, water vapour and radioactive particles over the Trinity Site bore the "spores" for the nuclearization of the world. It made worldwidely thinkable (and feasible) that the ancient millenarianist idea of the total annihilation of human life on Earth could be fleetingly achieved by technological means. It is not by chance that the year 1945 CE (Common Era) has been underlined by some scientists as the year 0 AE (Anthropocene Era). In the current Anthropocene debates, the start of a new geological stratum induced by human activity has been associated with the subsequent sociotechnological "revolutions" to control nature: the Agricultural Revolution (especially following the creation of the first agrarian states), the Scientific Revolution, the Industrial Revolution, the Green Revolution, and, finally, "the Atomic Revolution" (Lewis and Maslin 2015).

The world had already been *one* landscape for such different traditions for a long time: Christians and Muslims, political economists and socialists of every hue, esperantists and anti-colonial millenarists, and, of course, techno-optimists and techno-busters. The advent of railways, the telegraph, the telephone, radio, the aeroplane, TV and the WWW have – successively and amnesiacly – evangelized the coming of a merry, interdependent "global village" (Edgerton 2006, 105–117). At the same time, deforestation, steam-engine smoke and chemical products were also understood as seeds of catastrophic risks on a world scale, especially before being normalized through "little modern disinhibitions" which have led towards our current "joyful apocalypse", as Jean-Baptiste Fressoz has pointed out (Fressoz 2012, 9–25; see specifically Grove, 1995, 168–388, about colonial deforestation and "global environmentalism").

The appearance of the nuclear mushroom thus did not lead the world to be "as one" for the first time, nor was it the starting point of the globalization of technological risks. However, during the Cold War, nuclear technologies – along with the development of sciences such as space sciences, geosciences and ecology – paved the way for new ideas on the "global environment" (Grevsmühl 2014; Camprubí 2016). Moreover, nuclear technologies provided technological risks and potential catastrophe with a new spacial and temporary dimension. According to philosopher Timothy Morton, nuclear "hyper-objects" such as A-bombs and H-bombs have thrown a new veil under the Sun for the next centuries and millennia due to its ubiquity, speed and transcendence: with them, the end of the world has still come (Morton 2013).

Is it a new epoch, here and now? Since the prefix of Anthropocene has caused unease in many people ("which *anthropoi*? which humans? indifferently?"), a number of alternatives to the term have recently emerged: Androcene-Christocene-Capitalocene-Plantationocene-Corporatocene-Chthulucene-Thermocene-Phagocene-Tanathocene-Agnotocene-Polemocene-Elachistocene-Plasticene [...] (Schneiderman 2015; Moore 2016; Bonneuil and Fressoz 2017). Many other terms could be added to the list (in the event that they have not been invented and discussed prior to this publication). For instance, we could add Statocene and Highmodernocene if we were to draw upon the work about the state and the high-modernist ideology by anarchist-leaning scholar James C. Scott (Scott 1998, 2012). Or, a term which is not at odds with the last two: Uraniocene?

The Forbidden Fruit in a Globescape

#PLAY TRACK. Bradley Kincaid, "Brush The Dust From That Old Bible" (1950) [02:34] + Pink Floyd, "Come In Number 51, Your Time Is Up" from the final scene of *Zabriskie Point* (Michelangelo Antonioni, 1970) [06:56] + Los Ganglios, "Hay", LP *La guapa y los ninjas* (2002) [04:51]#

Once upon a time, technology had been disturbing the beauty, pureness and silence of what were perceived as pastoral paradises, as Leo Marx said in *The Machine in the Garden* (Marx 2000). Not just in idyllic nature, of course: the uneasiness about and resistances to machines by King Ludd and Captain Swing started at the very moment that they entered the factory and the farm.

When the first world economic crisis broke out in 1929, the machine was firmly put under the spotlight and under suspicion. Not just for its material responsibility, but also for other, deeper reasons. Philosopher José Ortega y Gasset fiercely criticized the technological processes of dehumanization, and lambasted the mass-man for believing that technology was a natural fruit of the Garden of Eden (Ortega 1932 [1929], 82, quoted in Marx 2000, 7–8; Mitcham 2005, 950–952):

The world is a civilized one, its inhabitant is not. [...] The new man wants his motor-car, and enjoys it, but he believes that it is the spontaneous fruit of an Edenic tree. [...] [He] does not extend his enthusiasm for the instruments to the principles which make them possible.

Ortega y Gasset could surely grasp the popular techno-enthusiasm, for example, during his visit to the 1929 International Exhibition in Barcelona, which became a huge showcase of engineering display and technological sublime like any other world fair. However, the social debates on the "machinery question", namely on technological unemployment, on chemical weapons, and on industrial democracy and technocracy became increasingly heated over the years, especially following the financial crash. Besides, Ortega seemed to turn a blind eye – see the quote above – to the great popularity of "pure sciences" and their principles, such as astronomy and relativistic physics (Roca-Rosell and Ruiz-Castell 2016; Glick 1988). Ortega was much more given to recognizing a sinless new Adam in Albert Einstein than in the prosaic engineer or the mass-production worker (Ortega 1923). Even so, he could not imagine that the "innocence" and "pureness" of this physicist would be so relevant in the race of providing a new "forbidden fruit" to the world in 1945, as Lewis Mumford suggested some decades later in *The Myth of the Machine* (Ortega 1923; Mumford 1970, 255).

At the end of World War II, the shadow of the mushroom clouds covered all the creatures and critters of the world. It seemed that a new fifth rider of the Apocalypse – leading the other biblical riders of conquest, of war, of famine and of death – could show up at any moment and destroy any blade of grass. In September 1945, the first foreign journalist who visited the horrific landscape of Hiroshima wrote in the newspaper article "The Atomic Plague" (Burchett 1945):

> Hiroshima does not look like a bombed city. It looks as if a monster steamroller had passed over it and squashed it out of existence. I write these facts as dispassionately as I can in the hope that they will act as a warning to the world.

He was not the first to notice that the destructive effects of nuclear technology could acquire a global dimension and could make planet Earth into a single death landscape or tanathoscape. Some weeks before the final decision to target Hiroshima and Nagasaki, Leo Szilard and nearly 70 scientists involved in the atomic project tried to prevent the US president Harry Truman from dropping the bomb. In the famous petition to the president they highlighted the possibility of "opening the door to an era of devastation on an unimaginable scale" and of providing "almost no limit to the destructive power" (Szilard 1945). Even Truman himself also depicted the Trinity nuclear test on 16 July as "the fire destruction prophesied in the Euphrates Valley Era, after Noah and his fabulous Ark", as Peter Kuznick has reminded us (Kuznick 2007, 1). "I am become Death, the shatterer of worlds!" was also proclaimed by J. Robert Oppenheimer, in charge of the secret Los Alamos Laboratory. Even some years before the first nuclear conflagration, scientists and bureaucrats behind the Manhattan Project seemed to be aware that a new turning (and not returning) point could be "knockin' on Earth's door". In 1942, the head of nuclear research at the Metallurgical Laboratory in Chicago, Arthur Holly Compton, confided his restlessness to Oppenheimer in the following terms: "better to accept the slavery of the Nazis than to run a chance of drawing the final curtain on mankind" (Compton 1956, 128; Kuznick 2007).

Going back to the first years of the 1929 crisis, psychoanalyst Sigmund Freud warned about the perils of civilization (and nature) in a kind of revised lyric of the Song of the Sibyl (Freud, 1962 [1930], 92):

> Men have gained control over the forces of nature to such an extent that with their help they would have no difficulty in exterminating one another to the last man. They know this, and hence comes a large part of their current unrest, their unhappiness and their mood of anxiety.

The Gospel of Nuclear Edens

> #PLAY TRACK. Opening theme of *Dr. Strangelove or: How I Learned to Stop Worrying and Love the Bomb* (Stanley Kubrick, 1964) [03:20] + Elton Britt, "Uranium Fever" (1955) [02:12] + Lolita Sevilla, "Americanos" from *Welcome Mr. Marshall!* (Luis García Berlanga, 1953) [02:09]#

Some years after the "atomic plague" was heralded, a new Edenic Garden was promised: a global techno-paradise run by atoms.[2] Scientists and scientific popularizers such as Frederick Soddy and Muriel Howorth defended early on that "the nation which can transmute matter could transform a desert continent, thaw the frozen poles, and make the whole world one smiling Garden of Eden" (quoted in Johnson 2012, 553, 567). In particular, irradiated seeds and radioactive plants were supposed to renew the sap of both the "tree of knowledge" and the "tree of life". Biblical metaphors were extensively used to describe a new, united, brave world, as in Howorth's book *Atom and Eve* (1955).

In 1953, the Atoms for Peace programme was launched by US president Dwight D. Eisenhower in the United Nations General Assembly. The programme announced a high-modernist global wonderland, the abundance of which was to be provided through scientific expertise, irradiated crops and "too cheap to meter" energy from fission reactions in nuclear power plants (Forgan 2003, 188–191). Since this moment, nuclear rhetorics – alongside nuclear things and nuclear energy – circulated as rapidly as the Atom Ant, from the dirty and hazardous mines in Portugal, Namibia and the Navajo Nation to the warm and comfortable living rooms in Paris, Tokyo and New York (concerning the mines, see respectively Marinho 2002; Hecht 2012; Gilles 1996).

During the Cold War, nuclear power was extolled as a symbol of international peace and radiant modernization of all humankind – all this despite its key role in the heated struggle for geopolitical hegemony in the postcolonial world and the dramatic increase in the number of warheads (from 1953 to 1986, the world number of warheads increased from 1,290 to 69,368 (98.5 per cent in the hands of the USA and USSR) (Norris and Kristensen 2010). Hand in hand with the proliferation of nuclear national programmes, nuclear internationalism and nuclear globalism spread all over the world (Krige 2006; Edgerton 2007): from the Far East to the Far West, from communist Eastern Europe to fascist Southern Europe, etc. The International Atomic Energy Agency (established in 1957), borrowed the flag of the United Nations and replaced the original image of a world map with the image of an atom – in fact, the image of a three-electron atom in its first version; that is, a lithium atom, the very coveted fuel for hydrogen bombs.

Especially since the International Conference on the Peaceful Uses of Atomic Energy held in Geneva in 1955, a huge amount of money, resources, films and exhibitions were invested to make the promised nuclear paradise desirable for politicians and citizens of the superpowers and the so-called "power-starved countries" (Krige 2010). The "gospel" of atomic energy was foretold from New Delhi and Tehran to Accra and Madrid, from Ceylon and Malaysia to Venezuela and Korea, and of course at the core of the Japanese apocalyptic nightmare (Yuka 2014).

In Hiroshima, nearly a million people – many of them being a "captive public" as school students – visited the travelling exhibition *The Peaceful Uses of Nuclear Energy* in the A-bomb Museum between 1956 and 1958. There, visitors were urged, for instance, to write the words *"heiwa"* and *"genshi ryoku"* (peace and nuclear energy in Japanese) with the mechanical arms used to manage uranium in specialized laboratories (Zwigenberg 2012). Supported by the US Department of Defense as well as by the city council, the regional prefecture, universities and local newspapers, this exhibition played a crucial role in facing the huge popular resistance in Japan to nuclear proliferation. In this country, millions of signatures were collected after March 1954, when a US nuclear test in the Bikini Atoll led to the contamination of the tuna fishing boat Daigo Fukuryu-Maru (Lucky Dragon 5), the poisoning of the fishermen, the panic in the fish markets and the extensive mobilization of housewives (Tanaka 2011). In the context of Nippon as in many others for decades, technological displays, entertainment and fun became critical tools for rendering the nuclear landscape banal and natural (Sastre-Juan and Valentines-Álvarez forthcoming).

Even more disturbingly, not only were atomic seeds and power plants supposed to provide a healing and lush Garden but nuclear weapons were also to be a source of life. The post-World War II official "triumphal narrative" transmuted the atomic bombs of the Armageddon – which had been called "Little Boy" and "Fat Man", and had killed 200,000 people in two blinks of an eye in Hiroshima and Nagasaki – into a kind of new "Saviour" (Dower 1997). According to this narrative, these bombs had supposedly accelerated the end of the war, "saving" millions of Western bodies and souls. During the Cold War, governments, armies and scientists also contended that the reliability of a huge nuclear arsenal – which was waiting for the final hotline call – was the reason for not having a nuclear conflict of unimaginable dimensions (Gusterson 2004, 151). Besides, a number of ritual metaphors of life and birth were associated with nuclear weapons (or with the scientific processes to design them). The first unshielded nuclear reactor in Los Alamos, for example, was known as the mythical "Lady Godiva" (Godgifu, Gift of God, in Old English). Anthropologist Hugh Gusterson has recalled a number of these metaphors: "little boy", "daughter", "babies", "cradle", "crib", "father", "breeder", "umbilical cords", "marriage", "generations", etc. In Gusterson's terms (Gusterson 2004, 161–163),

> [I]n metaphorically assimilating weapons and components of weapons to a world of babies, births, and breeding, weapons scientists use the connotative power of words to produce – and be produced by – a cosmological world where nuclear weapons tests symbolize not despair, destruction, and death but hope, renewal, and life.

#PLAY SOUNDS. Beeps in *1945–1998* (Isao Hashimoto 2003) [14:24] .
[overlapping] Birdsong by Horne Lark (*Eremophila alpestris*) [01:54] +
Cactus Wren (*Campylorhynchus brunneicapillus*) [02:11] + Common Raven
(Corvus corax) [01:05]#

Throughout the second half of the twentieth century, the world was exponen-
tially targeted by nuclear weapons, as video artist Isao Hashimoto has made
audible in the time-lapse map artwork *1945–1998* (Hashimoto 2003). One of
the most badly bombed places was the Nevada Test Site. The 3,500-km² test
site in south-central Nevada is a harsh and dry transitional zone between the
Mojave desert and the Great Basin desert. There, atomic tests made "natural
landscapes" (once inhabited by indigenous peoples) into nuclear landscapes
(surrounded by radiation-exposed indigenous peoples) (Russ *et al.* 2004). From
1951 to 1992, hundreds of nuclear bombs from the US Army left this vast salt
flat pockmarked with 300-metre lunar craters. The nuclear site became an
extreme representative of the confluence of conflict, environment and technology on planet Earth (LoPresti 2016a). As the scientific pictures as much as the
art pictures by Peter Goin seem to suggest, the site can be imagined as a por-
trayal of the apocalyptic ruins (Goin 1991).

In the 1960s, artists such as Michael Heizer created works amidst the militarized and nuclearized US west by channelling latent feelings of political impo-
tence into the largest and most physical gestures available to them, a movement
later called "Land Art" (LoPresti 2016a; Celant 1997, 203–289). Perhaps
because of the technological awe experienced by these artists, this movement
generally failed to take into account what survived the nuclear blasts and their
radioactive after-effects. Among the "survivors" we find fungi, such as the
Podaxis pistillarisfungi, popularly known as the Desert shaggy mane mushroom.
But we also find resilient and tough species from the plant kingdom, such as the
Fishhook cactus, the Kaibab agave, the Shinyleaf sandpaper, the Purple sage and
the Apache plume, all with their fragile and colourful flowers. At the same time,
the Rattlesnake pit viper, the Cooper tortoise, the Chuckwalla iguana and other
underground, reptilian and venomous critters once lived on the debris of one of
the most nuclearized spaces in the world (Tanner and Jorgensen 1963).

In contrast, recent artists such as Eric LoPresti have made life survival
their explicit subject. This artist grew up near the Hanford site, affording
him an unusually close relationship with nuclear weapons production. This
familiarity with militarized US deserts profoundly shaped his ways of seeing
and painting. LoPresti's recent series of large-format watercolour paintings
juxtapose the blasted landscape of the Nevada Test Site with paintings of
Bitterroot flowers (*Lewisia rediviva*), a resilient American shrub which grows
in and adjacent to the site. Whatever apocalyptic unease remains in LoPresti's
work, it does not prevent the observation that, on an ecological scale, both
nuclear craters and ephemeral flowers are contingent upon receptive viewers
for their meaning. In his words, both are "meant to be observed" (LoPresti
2016b).

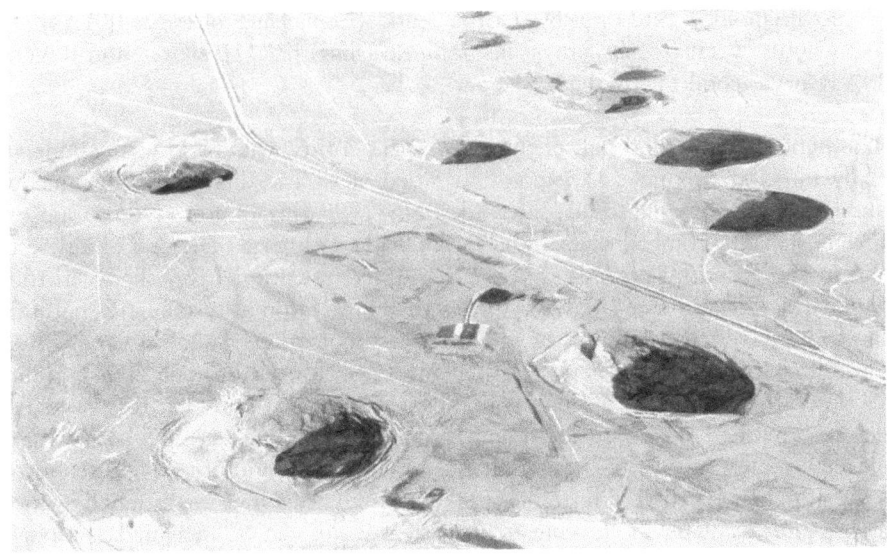

Figure 10.1 *Dark Red and Blue Craters*, watercolour on paper, 50′38″ (2016). A colour version of this figure is available to view online at https://www.routledge.com/9780815346661

Source: Artist's own image. © Eric LoPresti.

Figure 10.2 *Pink Lewisias*, watercolour on paper, 50′38″ (2016). A colour version of this figure is available to view online at https://www.routledge.com/97808 15346661

Source: Artist's own image. © Eric LoPresti.

Neither did the birds (e.g. the Horne lark, the Cactus wren or the Common raven) stop singing (Hayward, Killpack and Richards 1963).

Utopians are Everywhere

#PLAY TRACK (a heterogeneous Iberian "No Nukes" collectanea). Peret, "Yo soy gitano" (1972, Catalan rumba) [03:24] + Fausto, "Rosalinda", LP *Madrugada dos Trapeiros* (1979, protest song) [02:58] + La Bullonera, "El verrugón atómico", LP *La Bullonera, III* (1979, folk) + Las Vulpes, "Central nuclear", LP *Quiero ser una zorra* (1983, feminine punk) [01:53] + Chicho Sánchez Ferlosio and Rosa Jiménez, "Encuesta junto a una central nuclear" ([1980–1985], vaudeville) [02:41] + Lluís Maria Panyella and Toni Giménez, "I va fer un pet" (1986, children's song) [01:13]#

At the beginning of the 1960s, however, it seemed that birds were singing less and less. For most people all over the globe, the nuclear Eden – with its promises of abundance – appeared not to materialize. Nor did the alleged "man's progress not merely in knowledge but in the civility of his life on Earth", as Oppenheimer foretold at that time – he was, in fact, trying to "redeem" himself and his collaborators in Los Alamos Laboratory by fostering the discourse that linked the scientific revolution with the democratic institutions, a discourse which became essential in Cold War ideology (Oppenheimer 1984 [1963], 142).

For many people, the never-coming Eden was even turning into a quiet Hell. In 1962, biologist Rachel Carson was missing the sound of the birds in what seemed to be a "silent spring", and warned against the terrible daily effects of chemical products and radioactivity stemming from weapons, reactors and laboratories. She put into words a social anxiety which was to grow stronger and stronger (Carson 1994 [1962], 5–6):

> Only within the moment of time represented by the present century has one species – man – acquired significant power to alter the nature of his world. [...] The chain of evil [pollution] initiates not only in the world that must support life but in living tissues is for the most part irreversible. [...] Strontium 90, released through nuclear explosions into the air, comes to earth in rain or drifts down as fallout, lodges in soil, enters into the grass or corn or wheat grown there, and in time takes up its abode in the bones of a human being, there to remain until his death.

Carson had a great influence on the ecologist and anti-nuclear movements that arose globally since the late 1960s. Like the pro-nuclear programme of Atoms for Peace, the anti-nuclear movement had a remarkable transnational dimension (although of a different type, since it was essentially bottom-up and decentralized). The most visible proof is perhaps the "Smiling Sun" – with the leitmotiv "Nuclear Power? No Thanks" – which smiled in dozens of languages on millions of badges and stickers in all the continents. Anti-nuclear scientific reports, magazines, leaflets, news, actions and people circulated across borders. Former NIMBY resistances (acronym for "Not In My Back Yard") gave way to NIABY opposition to nuclear technologies (acronym for "Not In Anyone's Backyard"). According to the so-called "anti-nukes", the framework of resistance had to be worldwide so long as radioactivity was neither a nail nor a snail: it could not be attached and it travelled quickly.

Shortly after the nuclear meltdown in Chernobyl, philosopher Günther Anders wrote: "Chernobyl is everywhere" (Anders 2013 [1986]). In fact, he added this sentence to the previous "Hiroshima is everywhere". With these statements, he sought to sum up the idea that any place in the world could be targeted by the bomb, and every place in the world was put in danger by radiation. After having seen a beam of hopefulness in the dramatic psychological disorders of the "Hiroshima pilot" Claude Eatherly (with whom Anders kept a touching correspondence), he vindicated the state of panic about the possibility of a global genocide, or, in his terms, the possibility of a "globocide" (Anders 1961, 2013). During the last years of his life, these conclusions drove Anders to move from promoting pacifist practices of resistance to opening up the possibility of violent direct action, since, in legal terms, people would be living in a "state of necessity" (Anders 2008 [1987]). In fact, nuclear violence was not just a real or potential consequence of energy programmes, as was shown in Mayak, Sellafield, Idaho Falls, Lucens, Madrid, Jaslovské Bohunice, Harrisburg, Tsuruga, Chernobyl and so on (the estimated number of premature deaths in Chernobyl ranges from several thousands to nearly a million). Violence was also an actual means of their production: as in the conquest of the Far West, the states and the private companies used barbed wire and guns to found the nuclear Eden. Among the many arrested, imprisoned and injured protesters by police nightsticks, bullets and tear gas grenades, physics teacher Vital Michalon in Malville (France) and chemistry student Gladys del Estal in Tudela (Spain) were killed in pacifist anti-nuclear demonstrations in 1977 and 1979 (Tompkins 2016, 158–193; Lemoiz 1987).

Besides criticisms of (and actions against) the dreadful and violent features of the nuclear threat, the anti-nuclear movement constantly expressed their willingness "for life" and "good places". They asked for (and developed) playful work, sustainable communities, renewable energy, the recovery of abandoned villages, squatting, organic agriculture, bioarchitecture, and the use of bikes and plenty of gadgets which took advantage of the energy of atomic reactions produced millions of kilometres away in the Sun: solar ovens, thermal collectors, wind turbines, wave power turbines, domestic biogas plants and other so-called "appropriate technologies", "soft-technologies" or "liberatory technologies".

There was no place for single, universal, abstract, individually designed, masculine, unblemished and Edenic utopias. In this way, many "anti-nukes", radical ecologists and anarchists not only rejected the nuclear Eden but also most of the former dreamed (and usually scientifically managed) landscapes of famous utopians and utopian socialists, from Thomas More, Charles Fourier and

Étienne Cabet to Dwight Eisenhower. In *History and Utopia* (1960), E.M. Cioran cynically stated (Cioran 2015 [1960], 85):

> In Fourier's "societary state", they are so pure that they are utterly unaware of the temptation to steal, to "pick an apple off a tree". But a child who does not steal is not a child. What is the use of creating a society of marionettes?

However, a tree without any utopia may seem to be completely dried up. In a previous and more detailed critique of utopias, Lewis Mumford also made this point (Mumford 2015 [1922], 279–280):

> The weakness of the utopian thinkers consisted in the assumption that the dreams and projects of any single man might be realized in society at large. [...] Where the critics of the utopian method were, I believe, wrong was in holding that the business of projecting prouder worlds was a futile and foot-ling pastime. These anti-utopian critics overlooked the fact that one of the main factors that condition any future are the attitudes and beliefs which people have in relation to that future.

To collectively look for and find new utopias (in the plural, with apples to be stolen and small technologies to be self-managed) and to develop and take care of *eutopias*, "good places", seemed to be of great interest in facing the nuclear Eden. "By striving to do the impossible, man has always achieved what is possible. Those who have cautiously done no more than they believed possible have never taken a single step forward", Mikhail Bakunin stated shortly before the Paris Commune in 1871.

> #PLAY SOUNDS. West Coast, *Cuckoo Clocks* (2017) [02:58] and *Bird Sampling* (2017) [02:28]#

Mushrooms may represent death and technological violence, but they are a spontaneous and delicious food from the wilderness and beyond. During the Cold War, they were especially appreciated by the self-organized and self-sufficient communities of anti-nuclear hippies, punks and ex-yuppies who decided to flee the city for the countryside and the forests. Moreover, mush-rooms revealed the critical relevance of preserving traditional knowledge and communal know-how for survival, since this knowledge and know-how was essential to avoid serious poisoning from toxic and lethal species. Some tradi-tionally used toxic species of mushrooms, nonetheless, could be greatly wel-comed at certain moments: for many counter-culture youngsters, Psilocybes and other hallucinogenic mushrooms could become the most direct way to be "integrated" into the natural and supernatural worlds (Schultes, Hofmann and Rätsch 1979). "Drink me", Alice reads on the bottle before she embarks on a strange travel experience with surrealistic shifts of time and space during her adventures in Wonderland. Curiously, Lewis Carroll's book had been extensively used during the 1940s and 1950s to popularize relativistic physics and to promote advanced technology as a liberating force for mankind (Forgan 2003).

In the 1970s and 1980s, an ephemeral counter-culture art flourished in publications, journals of radical ecology, anarchist fanzines, underground comics, books of popularization, and on badges and stickers. Along with other artistic expressions such as theatre, street installations, performances and music employing all kinds of rhythms (from protest song to feminist punk and children's songs), drawings, graffiti and collages depicted gigantic apocalyptic scenarios while making fun of the official myths associated with the nuclear garden (Valentines-Álvarez and Macaya-Andrés in preparation). Collages (for example, the photo montage *Nuclear Enchantment*, by artist Patrick Nagatani, on the commemorative obelisk at the Trinity Test Site) are still a source of confronting official history and destabilizing narratives of the past (Masco 2006a).

The recent sound artworks *Towards a Libertarian Technology*, by Francisco Pinheiro, and *Cuckoo Clocks*, by West Coast collective, seem to represent as much as to pursue the spirit of these "ways of doing". West Coast seeks to be a nomadic platform of creation and debate on coastal cultures, science and human ecology, which aims to encourage different social actors such as academic researchers, beekeepers or watchmakers to work together collaboratively (West Coast 2017). After thinking, crafting and testing during many days on the patio of Marquis of Pombal Palace in Lisbon and in the ancient convent of Montemor-o-Novo (Portugal), Pinheiro, Paulo Morais and other members of West Coast – being both artists and artisans – have shown how to make a waterwheel turn in order to ring a little old bell and to turn time around, or how to (re)produce bird songs from organic elements and waste materials gleaned from post-industrial landscapes in order to create a space of awareness for endangered species (West Coast 2017; Pinheiro 2017, 84–105).

Updating: Nuclear_Eden.02

> #PLAY TRACK. "The Garden/In the Golden Afternoon" from Walt Disney's *Alice in Wonderland* (1951) [03:49] + "Always Look On The Bright Side Of Life" from the final scene of *Monty Python's Life of Brian* (Terry Jones, 1979) [03:22]#

Nowadays, a renewed nuclear Eden is being announced for the sake of all humankind. It is no longer the garden of Cornucopian abundance but a technocratically managed globescape for the sustainability of capitalism, in which there will be no biblical plagues, inundations, hurricanes, droughts, fires, rising sea levels or other "punishments for our sins" (those plagues that the poor in particular are experiencing more and more because of the consequences of global warming). Nuclear technology is urged to control the weather, to reverse global climate change and to overcome the lack of energy resources: eco-modernists, geoengineers, right libertarians, former environmental gurus and electrical corporations have strongly campaigned in favour of nuclear power for being a "green" and "sustainable" energy source (Lovelock 2004).[3] Westinghouse adver-tises on its website: "Nuclear energy is the largest source of clean electricity in the world. No *carbon* emissions, and no air pollution. Just safe, clean, and reli-able electricity" (Westinghouse 2018).

Turning the language upside-down, it is even said that the power of the atoms that can annihilate a town, a region or the whole of humankind is now the super-techno-fix needed for the final survival of humankind (Miller 2013). The nuclear promoters seem to have melted the former discourses of nuclear paradises which have spread since the 1950s with an apocalyptic language about "the end of nature" borrowed from Bill McKibben and other environmentalists. In the words of professor of ethics Michael Northcott, they somehow announce an "anthropic epiphany" through which humans beings would become the redeemer and the redeemed "for the healing of the nations", as if it was an updated version of John of Patmos' *Book of Revelation* (Northcott 2015, 104–106).

The new discourses are evangelized regardless of the already relatively long nuclear history of empty promises, experienced suffering and political domi-nance, and without mention of the interests of the military-industrial complex, the energy stock market, the business of technological surveillance, and the maintenance of the political and social order. The greatest impact of nuclear engineering might not have been to produce some artefacts that are able to annihilate the world, but also to produce everyday consequences in society and politics. As Joseph Masco suggests, a new kind of "secret state" has been estab-lished based on the policies of the "secret science", new forms of authority have been developed in democratic regimes, and nation-states have been reinforced by what he calls a permanent "nuclear state of emergency", in which govern-ments have legitimacy to do anything they like with regard to national (in)security (Masco 2006b). "The nuclear issue is not a technological or scient-ific issue; it is simply a social issue", Jaime Semprún wrote in *The Nuclearization of the World* in 1980. According to him, a state – in both terms of the word – of ignorance and control had been essential to the nuclear programme (Semprún 1982 [1980]). A decade previously, Lewis Mumford had been categoric: the "nuclear pyramid" was the last step of a technocratic and bureaucratic endeav-our that had started during the time of Ancient Egypt and epitomized "the uni-versal imposition of the megamachine" (Mumford 1970).

Science can methodologically and meticulously diagnose the catastrophe by providing plenty of data, percentages, graphics, diagrams, mathematical correla-tions, etc. (Debord 2015). Nevertheless, due to its limits, the dominant values it mirrors and the fact that it is a significant cause of the problem, science – espe-cially nuclear science – cannot provide tools to face the main sociotechnologi-cal challenges (Wynne 1992). It is necessary to do it by other means. And so the story goes on and on.

Finale? The Apocalyptic Fungus (II The Mycelium)

#PLAY TRACK. Coetus, "El cant de la Sibil·la [Song of the Sibyl]" (2009) [10:12]#

A mushroom is just the ephemeral fruiting body of a much larger being whose vegetative part – called mycelium – can spread unnoticed some inches or many kilometres underground. A large, process-complex and branched mass of filaments called hyphae (from *huphe*, meaning "web" in Ancient Greek) is hidden behind the mushroom. These hyphae can develop extraordinarily deep associations with roots of plants and with other living beings, such as the symbiotic associations called mycorrhizas which play a critical role in the circulation of nutrients and water, and in the composition of the soil. In fact, some of these forms of mutualist relationships appeared when plants started to occupy the land hundreds of millions of years ago. After quietly waiting enough time for the appropriate atmospheric and ground conditions, joined mycelia can make a mushroom appear on the surface, which will produce millions and millions of "reproductive fruits" or spores.

The mushroom has been the image of the end of the world since 1945, but it can also be a metaphor of the beginning of a new world, as the Apocalypse can be interpreted. In some ways, the biblical phrase *"Destruam et aedificabo"* – which was recalled by Pierre-Joseph Proudhon in *The Philosophy of Misery* in 1847 – seems to make total sense. The mushroom not only represents death and catastrophe, but also life renewal and a turn-around (*revolutio* in Latin). Apart from bacteria and archaea, fungi were the first living beings to appear in the ground following some of the most dreadful nuclear accidents. In this sense, mushrooms allow us to consider the possibility of life over the ruins after (or during) the "apocalypse", and even to think – as Anne Tsing makes clear – about the possibility of (re)placing biological and technological refuges for life through multispecies collaboration and mutual aid (Tsing 2015, 1–9; Dighton, Tugay and Zhdanova 2008). "Make kin, not babies!", Donna Haraway exclaimed in six of the most poetic pages the current "academia-industrial complex" has probably produced about the Anthropocene debates during the past years (Haraway 2015): make kin, kind, care, babies, parents, aunts, uncles, cousins, grandmas, regardless of blood kin.

In *Mutual Aid: A Factor of Evolution* (1902), Pyotr Kropotkin defended that evolution in the animal kingdom could not be just explained through struggle and competition for food and life, as most Darwinists defended. The fittest were not always the strongest nor the most cunning (Kropotkin 2013 [1902]). Evolution also depended on multiple kinds of association, cooperation among individuals within a species, and mutualism among individuals of different species. These relationships had to do with the need to overcome the "natural checks to over-multiplication", which could be of much greater importance in evolution than competition: in Eurasia, for example, animals have to face terrible frosts and snowstorms during the winter, and torrential rains and floods during the summer every year.

As urban geographer Mike Davis has recently reminded us, Kropotkin was also a renowned geologist who argued contemporaneously that we are living in a period of harsh desertification in the Holocene, that this global climate change is an outstanding mover of the history of mankind, and that people have to find the means to curb this situation (Davis 2016). According to Kropotkin, not only might we need to plant millions of trees and dig thousands of artesian wells to avoid the final "desiccation of Eurasia", but we also have to be aware that a wider extension of non-authoritarian relationships based on mutual aid is indispensable to life.

Many social experiences based on mutual aid and play-and-work collaboration have appeared in the midst of natural and human-induced catastrophes or over the "ruins" of once symbols of flourishing capitalism. The social revolution in Barcelona in the midst of the battle against fascism during the Spanish Civil War readily comes to mind, along with its plethora of anarchist and collectivist proposals put into practice. But we can also find many examples at the beginning of the twenty-first century: for instance, the collective action and sense of community among the ghostly architectures and the urban decay in Detroit; the self-management of workshops and factories following the financial *corralito* in Buenos Aires; the establishment of the so-called "democracy without state" during the war against ISIS in Kurdish Rojava; the self-organization of provisions and the non-state and leaderless system of emergency aid in the Caribbean Islands, Florida and Mexico DF which followed hurricane Irma and the earthquakes in 2017; or the solidarity-based work of the forestation of local tree species and the rebuilding of houses and communities after the devastating fires in Portugal in 2017 and 2018 (Gomes *et al.* 2018).

But do all these experiences really pop up like mushrooms in catastrophic landscapes? Absolutely; they do as mushrooms do: they just grow if (and only if) there is a previous dense web of invisible collaborative filaments from which dead matter can be recycled to become living matter again. It is not a sufficient condition but a *sine qua non*. To preserve, reinforce and take care of these mycellia is a pleasant duty (and perhaps the opportunity for a humble transcendence while expecting that delicious fruiting bodies will emerge in future generations).

Obviously, cooperative, self-managed and non-authoritarian experiences do not necessarily like living on debris. Fungi and other beings do not prefer the living conditions in Chernobyl's surroundings either: studies have demonstrated that their activity is slower than usual (Mousseau *et al.* 2014). Imagine, then, how vigorously and vividly they would grow in material conditions other than the worst dramatic and despairing circumstances.

Nuclear paradises, technoscientific command and voluntary servitude? "No thanks", the anti-nuclear movement would politely respond.

Energy sovereignty of individuals and peoples, technological self-management, and non-authoritarian uses of the "machine"? "Yes thanks!", reply millions of peasants, workers and indigenous communities from large web-based organizations such as Via Campesina and MOCASE, and by tiny "techno-political" groups based on autonomy such as Fem-hi-Gas (Via Campesina 2017; Fem-hi-Gas 2006).[4]

We
(we, maybe not all the critters on the Earth nor the whole humankind, but
quite large underground mycelia of individuals and collectives from all over
the world)
do
not
want
a techno-Eden anymore:
we want collectively-made and non-hierarchically-based utopias alive!

Acknowledgements

FCT MCTES – Project PTDC/IVC-HFC/6789/2014 – ANTHROPOLANDS – *Engineering the Anthropocene: The role of Colonial Science, Technology and Medicine on Changing of the African Landscape* FCT MCTES – Project PEst-OE/HIS/UI0286/2014.

Notes

1 A first draft of this chapter was presented – with a mushroom literally "on the table" – in the round-table "Landscape Fukushima. Dialogues on Hybrid Natures", organized by Laura Valls at the CSIC, Barcelona, on 16 June 2014. Since then, constructive comments, suggestions and criticisms from many researchers, colleagues and friends have made this text a more collaborative (and interesting) work. In this sense, I am especially grateful to Álvaro Fonseca, Ana Macaya, Claudia Guerrero, Ferran Aragon, Francisco Pinheiro, Gloria Domínguez, Inês Ponte, Ivo Louro, Jaume Sastre, Leonor Valfigueira, Leonor Vera, Lucia Tinghi, Luísa Sousa, Marta Macedo, Pedro Morais, Pedro Mota, Pepe Pardo, and to the communities of Voltors, Pandores, Puris in Barcelona, Boesgers in Lisbon and Chien Bacou in Lausanne.
2 For a discussion about former colonial techno-Edens see Grove 1995; Fiege 1999.
3 For critical insights into eco-modernism and right libertarianism, which defends capitalism without the state, see Hamilton 2013, 107–137; Ippolita 2017, 155–166.
4 We borrow the term "techno-politics" from Gabrielle Hecht, who uses it to refer to "the strategic practice of designing and using technology to constitute, embody, or enact political goals", even though we do not use it here in terms of national political goals (Hecht 2001, 256).

Bibliography

Anders, Günther. 1961. *Burning Conscience. The Case of the Hiroshima Pilot, Claude Eatherly*. New York: Monthly Review Press.
Anders, Günther. 2008 [1987]. *Estado de necesidad y legítima defensa. Violencia, sí o no*. Madrid: Centro de Documentación Crítica (translated from *Gewalt: ja oder nein, Eine notwendige diskussion*. Munich: Knaur).
Anders, Günther. 2013 [1986]. "Diez tesis sobre Chernóbil. Mensaje amistoso al Sexto Congreso Internacional de Médicos por el impedimento de una guerra nuclear". *Argelaga* 2: 47–50 (translated from *Tageszeitung*, 10 June 1986).
Bonneuil, Christophe and Jean-Baptiste Fressoz. 2017. *The Shock of the Anthropocene. The Earth, History, and Us*. London; New York: Verso.

Burchett, Wilfred. 2007 [1945]. "The Atomic Plague". In *Rebel Journalism: The Writings of Wilfred Burchett*, edited by George Burchett and Nick Shimmin (pp. 2–5). Cambridge: Cambridge University Press.

Camprubí, Lino. 2016. "The Invention of the Global Environment". *Historical Studies in the Natural Sciences* 46(2): 243–251.

Carson, Rachel. 1994 [1962]. *Silent Spring*. Boston, MA: Houghton Mifflin.

Celant, Germano. 1997. *Michael Heizer*. Milan: Fondazione Prada.

Cioran, E.M. 2015 [1960]. *History and Utopia*. New York: Arcade [ebook].

Compton, Arthur Holly. 1956. *Atomic Quest: A Personal Narrative*. New York: Oxford University Press.

Davis, Mike. 2016. "The Coming Desert. Kropotkin, Mars and the Pulse of Asia". *New Left Review* 97: 23–43.

Debord, Guy. 2015. *O planeta doente* [A Sick Planet]. Lisbon: Letra Livre.

Dighton, John, Tatyana Tugay and Nelli Zhdanova. 2008. "Fungi and Ionizing Radiation from Radionuclides". *FEMS Microbiology Letters* 281(2): 109–120.

Dower, John W. 1997. "Triumphal and Tragic Narratives of the War in Asia". In *Living With the Bomb: American and Japanese Cultural Conflicts in the Nuclear Age*, edited by Laura E. Hein and Mark Selden (pp. 37–51). New York: M.E. Sharpe.

Edgerton, David. 2006. *The Shock of the Old: Technology and Global History since 1900*. London: Profile Books.

Edgerton, David. 2007. "The Contradictions of Techno-nationalism and Techno-globalism: A Historical Perspective". *New Global Studies* 1(1): 1–32.

Fem-hi-Gas. 2006. "'Carta de presentació.' Can Voltor, Collserola". *IV Jornadas de Pre-Okupación Rural, Kan Pasqual* (non-published and non-electronically available DIY fanzine).

Fiege, Mark. 1999. *Irrigated Eden: The Making of an Agricultural Landscape in the American West*. Seattle: University of Washington Press.

Figueras, Monserrat, Jordi Savall and La Capella Reial de Catalunya. 1996 [1988]. *El cant de la Sibil·la I. Catalunya*. Paris: Auvidis.

Forgan, Sophie. 2003. "Atoms in Wonderland". *History and Technology* 19(3): 177–196.

Fressoz, Jean-Baptiste. 2012. *L'apocalypse joyeuse. Une histoire du risque technologique*. Paris: Éditions du Seuil.

Freud, Sigmund. 1962 [1930]. *Civilization and its Discontents*. New York: W.W Norton.

Gilles, Cate. 1996. "No One Ever Told Us: Native Americans and the Great Uranium Experiment". In *Governing the Atom: The Politics of Risk*, edited by John Byrne and Steven M. Hoffman. New Brunswick: Transactions Publishers.

Glick, Thomas F. 1988. *Einstein in Spain. Relativity and the Recovery of Science*. Princeton, NJ: Princeton University Press.

Goin, Peter. 1991. *Nuclear Landscapes*. Baltimore, MD; London: Johns Hopkins University Press.

Gomes, João, José Carvalho, Filipe Nunes, *et al.* 2018. "Para lá do fumo" [monograph issue], *Mapa. Jornal de Informação Crítica* 20: 5–16.

Grevsmühl, Sebastian Vincent. 2014. *La terre vue d'en haut: L'invention de l'environnement global*. Paris: Seuil.

Grove, Richard. 1995. *Green Imperialism: Colonial Expansion, Tropical Island Edens, and the Origins of Environmentalism, 1600–1860*. Cambridge; New York: Cambridge University Press.

Gusterson, Hugh. 2004. *People of the Bomb. Portraits of America's Nuclear Complex*. Minneapolis: University of Minnesota Press.

Hamilton, Clive. 2013. *Earthmasters: The Dawn of the Age of Climate Engineering*. London: Yale University Press.

Hamilton, Clive. 2015. "Human Destiny in the Anthropocene". In *The Anthropocene and the Global Environmental Crisis: Rethinking Modernity in a New Epoch*, edited by Clive Hamilton, Christophe Bonneuil and François Gemenne (pp. 32–43). London; New York: Routledge.

Haraway, Donna. 2015. "Anthropocene, Capitalocene, Plantationocene, Chthulucene: Making Kin". *Environmental Humanities* 6: 159–165.

Hashimoto, Isao. 2003. *1945–1998*. www.youtube.com/watch?v=cjAqR1zICA0 (accessed 3 May 2017).

Hayward, C. Lynn, Merlin L. Killpack and Gerald L. Richards. 1963. "Birds of the Nevada Test Site". *Brigham Young University Science Bulletin. Biological Series* 3(1): 1–27.

Hecht, Gabrielle. 2001. "Technology, Politics, and National Identity in France". In *Technologies of Power*, edited by Michael Allen and Gabrielle Hecht (pp. 253–294). Cambridge, MA: MIT Press.

Hecht, Gabrielle. 2012. *Being Nuclear: Africans and the Global Uranium Trade*. Cambridge, MA: MIT Press.

Howorth, Muriel. 1955. *Atom and Eve*. London: New World Publications.

Ippolita. 2017. *Tecnologie del dominio. Lessico minimo per l'autodifesa digitale*. Milan: Meltemi.

Johnson, Paige. 2012. "Safeguarding the Atom: The Nuclear Enthusiasm of Muriel Howorth". *The British Journal for the History of Science* 45(4): 551–571.

Jorge, Ângelo. 2004 [1912]. *Irmânia. Novela Naturista*. Vila Nova de Famalicão: Quasi.

Krige, John. 2006. "Atoms for Peace, Scientific Internationalism, and Scientific Intelligence". *Osiris* 21: 161–181.

Krige. John. 2010. "Techno-utopian Dreams, Techno-political Realities: The Education of Desire for the Peaceful Atom". In *Utopia/dystopia. Conditions of Historical Possibility*, edited by Michael D. Gordin, Helen Tilley and Gyan Prakash (pp. 151–175). Princeton, NJ: Princeton University Press.

Kropotkin, Peter. 2013 [1902]. *Mutual Aid: A Factor of Evolution*. St Louis, MI: Dialectics.

Kuznick, Peter J. 2007. "The Decision to Risk the Future: Harry Truman, the Atomic Bomb and the Apocalyptic Narrative". *The Asia-Pacific Journal: Japan Focus* 5(7). https://apjjf.org/-Peter-J.-Kuznick/2479/article.html/ (accessed 3 May 2017).

Lemoiz Apurtu, 1972–1987. 1987. [s.l.]: Euskadiko Antinuklear eta Ekologistak (Spanish version in www.euskaletxeak.org/lemoiz).

Lewis, Simon L. and Mark A. Maslin. 2015. "Defining the Anthropocene". *Nature* 519(12 March): 171–180.

LoPresti, Eric. 2016a. "Land Art and the Nuclear Landscape". *Art F City*. Published 21 June. http://artfcity.com/2016/06/21/img-mgmt-land-art-and-the-nuclear-landscape.

LoPresti, Eric. 2016b. *Blooms* [paintings exhibited at the Elizabeth Houston Gallery in New York City, 2017]. www.ericlopresti.com/2016-blooms-public (accessed 3 May 2017).

Lovelock, James. 2004. "Nuclear Power is the Only Green Solution". *Independent*, 23 May. www.independent.co.uk/voices/commentators/james-lovelock-nuclear-power-is-the-only-green-solution-564446.html.

Marinho Falcão, José, Carlos Matias Dias and Paulo Jorge Nogueira. 2002. "Mortalidade por neoplasias malignas na população residente próximo de minas de urânio em Portugal". *Revista Portuguesa de Saúde Pública* 20(2): 35–51.

Marx, Leo. 2000 [1964]. *The Machine in the Garden. Technology and the Pastoral Ideal in America*. Oxford: Oxford University Press.

Masco, Joseph. 2006a. "5:29:45 AM". In *Museum Frictions: Public Cultures/Global Transformations*, edited by Ivan Karp, Corrine Kratz, Lynn Szwaja and Tomas Ybarra-Frausto (pp. 102–106). Durham, NC: Duke University Press.

Masco, Joseph. 2006b. *The Nuclear Borderlands. The Manhattan Project in Post-Cold War New Mexico*. Princeton, NJ: Princeton University Press.

Miller, John Dudley. 2013. "A False Fix for Climate Change". *Bulletin of Atomic Scientists*, 11 September. https://thebulletin.org/false-fix-climate-change.

Mitcham, Carl. 2005. *Encyclopedia of Science, Technology, and Ethics*. Detroit, MI: Thomson Gale.

Moore, Jason, ed. 2016. *Anthropocene or Capitalocene? Nature, History, and the Crisis of Capitalism*. Oakland, CA: PM Press.

Morton, Timothy. 2013. *Hyperobjects. Philosophy and Ecology after the End of the World*. Minneapolis: University of Minnesota Press.

Mousseau, Timothy A., Gennadi Milinevsky, Jane Kenney-Hunt and Anders Pape Møller. 2014. "Highly Reduced Mass Loss Rates and Increased Litter Layer in Radioactively Contaminated Areas". *Oecologia* 175: 429–437.

Mumford, Lewis. 2015 [1922]. *Historia de las utopías* [The Story of Utopias]. Madrid: Pepitas de Calabaza.

Mumford, Lewis. 1970. *The Myth of the Machine. The Pentagon of Power*. New York: Harcourt Brace Jovanovich.

Norris, Robert S. and Hans M. Kristensen. 2010. "Global Nuclear Weapons Inventories, 1945–2010". *Bulletin of Atomic Scientists* 66(4): 77–83.

Northcott, Michael. 2015. "Eschatology in the Anthropocene. From the *Chronos* of Deep Time to the *Kairos* of the Age of Humans". In *The Anthropocene and the Global Environmental Crisis: Rethinking Modernity in a New Epoch*, edited by Clive Hamilton, Christophe Bonneuil and François Gemenne (pp. 100–111). London; New York: Routledge.

Oppenheimer, J. Robert. 1984 [1963]. "The Scientific Revolution and its Effects on Democratic Institutions". In *Uncommon Sense*, edited by Gian-Carlo Rota, N. Metropolis and David Sharpy (pp. 141–146). Boston, MA: Birkhäuser.

Ortega y Gasset, José. 1923. *El tema de nuestro tiempo. El ocaso de las revoluciones. El sentido histórico de la teoría de Einstein*. Madrid: Calpe.

Ortega y Gasset, José. 1932 [1929]. *The Revolt of the Masses*. New York: W.W Norton.

Pinheiro, Francisco, ed. 2017. *Kamal*. Lisbon: Sistema Solar (Design Ilhas Estúdio).

Roca-Rosell, Antoni and Pedro Ruiz-Castell. 2016. "The Sky above the City: Observatories, Amateurs and Urban Astronomy". In *Barcelona: An Urban History of Science and Modernity, 1888–1929*, edited by Oliver Hochadel and Agustí Nieto-Galan (pp. 181–199). New York: Routledge.

Russ, Abel, Patricia George, Rob Goble, Stefano Crema, Chunling Liu and Dedee Sanchez. 2004. *Native American Exposure to Iodine-131 from Nuclear Weapons Testing in Nevada*. Worcester, MA: Clark University. www.clarku.edu/mtafund/prodlib/clark/round5/Iodine-131.pdf (accessed 3 May 2017).

Sastre-Juan, Jaume and Jaume Valentines-Álvarez, eds. (forthcoming in 2019). "Fun and Fear: The Banalization of Nuclear Technologies through Display". *Centaurus* (special issue).

Schneiderman, Jill S. 2015. "Naming the Anthropocene". *philoSOPHIA. A Journal of Continental Feminism* 5(2): 179–201.

Schultes, Richard Evans, Albert Hofmann and Christian Rätsch. 1979. *Plants of the Gods. Their Sacred, Healing, and Hallucinogenic Powers*. London: Hutchinson.

Scott, James C. 1998. *Seeing Like a State. How Certain Schemes to Improve the Human Condition Have Failed*. New Haven, CT: Yale University Press.

Scott, James C. 2012. *Two Cheers for Anarchism. Six Easy Pieces on Autonomy, Dignity, and Meaningful Work and Play*. Princeton, NJ: Princeton University Press.

Semprún, Jaime. 1982 [1980]. *A nuclearização do mundo*. Lisbon: Antígona.

Szilard, Leo. 1945. "A Petition to the President of the United States. July 17, 1945". www.dannen.com/decision/pet-gif.html (accessed 3 May 2017).

Tanaka, Yuki. 2011. "'The Peaceful Use of Nuclear Energy' and Hiroshima". *The Asia-Pacific Journal: Japan Focus* 9(18)(1). https://apjjf.org/2011/9/18/Yuki-Tanaka/3521/article.html (accessed 3 May 2017).

Tanner, Wilmer W. and Clive D. Jorgensen. 1963. "Reptiles of the Nevada Test Site". *Brigham Young University Science Bulletin. Biological Series* 3(3): 1–32.

Tompkins, Andrew S. 2016. *Better Active than Radioactive! Anti-nuclear Protest in 1970s France and West Germany*. Oxford: Oxford University Press.

Tsing, Anna Lowenhaupt. 2015. *The Mushroom at the End of the World: On the Possibility of Life in Capitalist Ruins*. Princeton, NJ: Princeton University Press.

Valentines-Álvarez, Jaume and Ana Macaya-Andrés (forthcoming in 2019). "Making Fun of the Atom: Humour and Pleasant Forms of Anti-nuclear Resistance in the Iberian Peninsula, 1974–1984". *Centaurus*.

Via Campesina. 2017. "The Solution to the Climate Crisis is in our Peasant Struggle for Food and Energy Sovereignty!" Published 20 October. https://viacampesina.org/en/solution-climate-crisis-peasant-struggle-food-energy-sovereignty.

West Coast. 2017. "Sights and Sound". http://west-coast.pt (accessed 3 May 2018).

Westinghouse. 2018. "Benefits of Nuclear Energy". www.westinghousenuclear.com/Why-Nuclear (accessed 3 May 2018).

Williams, Rosalind H. 2008. *Notes on the Underground: An Essay on Technology, Society and the Imagination*. Cambridge, MA: MIT Press.

Wynne, Brian. 1992. "Uncertainty and Environmental Learning: Reconceiving Science and Policy in the Preventive Paradigm". *Global Environmental Change* 2(2): 111–127.

Yuka, Tsuchiya. 2014. . "The Atoms for Peace USIS Films: Spreading the Gospel of the 'Blessing' of Atomic Energy in the Early Cold War Era". *International Journal of Korean History* 19(2): 107–135.

Zwigenberg, Ran. 2012. "'The Coming of a Second Sun': The 1956 Atoms for Peace Exhibit in Hiroshima and Japan's Embrace of Nuclear Power". *The Asia-Pacific Journal: Japan Focus* 10(6)(1). https://apjjf.org/2012/10/6/Ran-Zwigenberg/3685/article.html (accessed 3 May 2017).

11 Inhabitants

Image politics in ongoing climate crisis

Mariana Silva and Pedro Neves Marques

Introduction

Inhabitants is an online video channel focused on producing short-form video episodes for online distribution, founded by both of us in 2015. Its aim is to connect contemporary art with journalistic research. We have since published over 25 videos – under 10 minutes each – on different political themes, such as the Anthropocene, geoengineering, social justice, museology and post-colonial archives, as well as, more recently, deep-sea mining, among others. Each episode is released on the channel's website, as well as on Facebook, YouTube and Vimeo, and is available to be embedded on other websites (e.g. partner NGOs and museums, etc.). The videos may be streamed for free and are under a creative commons licence.

Inhabitants asks broadly what media literacy might mean in the age of social media and online video platforms, such as Vimeo and YouTube, which have given rise to a multiplicity of new short-form video genres designed to be shared by users.

The channel's title, *inhabitants*, references the eponymous film by Armenian director Artavazd Peleshian, in which archival footage, sourced from different years and origins, of stampeding mammals, flocks of birds and other animals are spliced together. As we state on the website, the film shows "looped footage invoking a symphonic pathos with the escape of elephants, flamingos, monkeys, and birds fleeing from the camera lens". Tellingly, the low-quality version of the film as found online, contaminated by pixels and drop frames, suggested to us a further reading of it "as a preamble to the mass extinctions that haunt contemporary regimes of visibility".[1]

Species extinction itself is yet to be the topic of an episode; however, we at inhabitants take to heart the notion that complex systems, such as climate change or the geological scale invoked by the Anthropocene, but also the infrastructure of the internet or of globalized financial markets, pose difficult questions of representation and image-making. The difficulty of representing climate change, for example, is reflected in myriad theoretical concepts, from Bruno Latour's "hybrids" and Michel Serres' "quasi-objects" to Timothy Morton's concept of "hyperobjects" – all attempts at visualizing either the connectedness of scientific events, experiments and decisions (Latour 1993; Serres 2007, 224) or the inhuman scale of the problem at stake (Morton 2013).

In our website statement the problem of representation is pitted against the film/video camera and its unavoidable perspectival depiction of the world. Similarly, in his *Perspective as Symbolic Form*, art historian Erwin Panofsky (1927) already identified a preoccupation which filmmaker Harun Farocki (Elsaesser 2014) would later develop in his essayistic documentaries: the relation between technology, as part of a system of representation, and ideology. Today, how does the increasing impact of such global large-scale systems on our human, but also non-human, lives reflect an imagetic-ideological regime, connecting video journalism (be it professional or citizen-led) and amateur video-making to television, cinema and the visual arts?

Thematically, none of the videos addressed in this chapter tackles this question directly. Methodologically, however, the use of animation or infographic visuals in some of the episodes serves to highlight how much video today is not camera-based, particularly in the sphere of short-form journalistic pieces, which can very well be termed a genre of online video. In this chapter, we will primarily foreground issues of representation and visibility in a selection of episodes published in the past three years: an episode about the emergence of geoengineering technologies as a possible way to mitigate climate change, and the economic interests behind it; a multi-episode web series entitled *Anthropocene Issue*, which documented an eponymous educational event held at the Haus of World Cultures in Berlin between 2014 and 2015; a web series on anti-oil extraction struggles entitled *For an Oil Free Future*, which reacts to attempts at oil drilling and fracking in Portugal; and our current five-part series on the prospects of deep sea mining worldwide, co-authored with art curator Margarida Mendes. Below, we offer an overview of each of these episodes and series, illustrating our concerns, methods and visual approaches.

Finally, the role of the garden is subsumed under the current blurring of modernity's long-standing division between nature and culture, which Bruno Latour, among others, identified as early as 1993 in his book *We Have Never Been Modern*, and which has provided us with a framework under which to understand the concept of the Anthropocene and the role of images and technology within it. The garden is the anthropogenic site that calls forth representations of nature as well as the human relation to its environment, in the West and elsewhere. Perhaps the role of the garden, understood as a refracting prism onto which notions of nature are projected, is best illustrated by a story. In his book *Mimesis and Alterity: A Particular History of the Senses*, anthropologist Michael Taussig (1993) recounts a passage from the diaries of surrealist writer Roger Caillois. Upon returning to Manaus after a period spent up the Amazon River, Caillois spots a French-style garden and witnesses a fellow passenger on his boat exclaim, while looking at the garden, "Ah, at last nature!" By this Taussig suggests that the chaos of the Amazon forest did not register as such to the Western eye. Rather, it was the ordered, geometric regularity of the French garden – with the colonial undertones which such an exclamation reveals – that best elucidated how, for the modern mind, the concept of nature was found somewhere between the sublime incommensurability of an untamed nature outside society (the Amazon forest) and the necessity of parsing out this "nature" into ever more detailed and tamed geometries and units (the garden as a space of applied science). If the question of modern nature was one of scale, as we suggest, how does the increasing relevance of large-scale complex systems reflect back on this division? And, more importantly for us as filmmakers, what power and role do images play in this division?

From postmodern images to social media clickbait

How do media, such as images circulating online, relate to the Anthropocene? The same year of Latour's French publication of *We Have Never Been Modern*, philosopher Jean Baudrillard (1991) published a controversial series of essays under the title *The Gulf War Did Not Take Place*. In these essays Baudrillard turns his analysis of simulacra to the context of the then-unfolding (first) Gulf War. He defends that atrocities against Iraqis are under way, but that the unequal nature of the US-led attack cannot be called war. For these events to be perceived as war, the media had to create the simulacra of war, for which the repetition of televised footage of bombings in newscast became a leitmotiv.

Video-sharing platforms were born well after the advent of the second Gulf War: Vimeo in 2004 and YouTube in 2005, while Facebook only added its video-posting option in 2009. Contrary to Baudrillard's analysis of the role of televised news in the first Gulf War, the change in medium from the television to the internet as a news source heralds a different phase of the making and dissemination of images: it is no longer human-led editorial guidelines that lead video clips to repeat *ad absurdum* in every news flash; rather, it is online viewing by video platform users that influences the visibility, the virality, of such video-recorded events. That is to say, algorithms prop up results in search engines and shares in social media. These algorithms are often opaque and not properly understood by users, and are infrastructurally tied to economics through paid ads and patents and proprietary rights.

Online media demand our participation in the circulation and mediation of images. In doing so, this involves each viewer or user in the visibility or invisibility of images. As is known, traditional news media have been dealt a hard blow by social media and other channels – Whatsapp, Wechat, etc. – and are now desperately trying to navigate this "democratized" space themselves. This has meant their subjugation to "virality" and "clickbait" content. On the subject of ecology, for example, while images of disaster such as the Deepwater Horizon oil spill in the Gulf of Mexico in 2010 evoke those of the Exxon Valdez oil spill of 1989, their mode of dissemination and repetition – what makes their images emerge from an otherwise overwhelming stream of images –is of a different nature. Images of Deepwater Horizon and other contemporary environmental disasters will increasingly depend on their visual capacity to engage viewers through clicks and shares in order to remain in the public eye. Traditional news media are increasingly being influenced by this logic, which then reflects back on their video editing choices and dissemination strategies.

On the other hand, protests such as the Dakota Access Pipeline or the NODAPL activist movement, organized by several Native American tribal nations (particularly the Lakota) at Standing Rock in the United States starting from early 2016, would not have become a worldwide phenomenon were it not for the savvy literacy of local onsite groups in using social media tools. Here as elsewhere, protests and their online "virality" ultimately ruptured mainstream televised media silence.

The production of images and the infrastructure of their circulation go hand in hand, shaping their form and content. Because of this, working between the visual arts and media, we at inhabitants look at how different video forms and genres are enacted online, from activist campaigning to academic lectures, citizen journalism to tutorials, from amateur video essays to video-game walk-throughs. In each case, their post-production techniques, animation and montage are intimately related to the rules of the internet. For example, title and text are often disassociated, and it is the former that will cause viewers to pause in their endless scrolling, click on contents and open a given video; or otherwise the video will be embedded directly on autoplay mode, with the scroll activating the video. These videos are structured by what their titles promise and the context at stake; between content and viewers' expectations; that is, viewers' attention span.

"A brief history of geoengineering:" climate markets vs. ontological debates

Inhabitants first episode, published in September 2015, is entitled *A Brief History of Geoengineering*. Geoengineering is described by the Oxford Geoengineering Programme (OGP) as "the deliberate large-scale intervention in the Earth's natural systems to counteract climate change". OGP's website goes on to describe its two main branches of research: Solar Radiation Management, which aims to "reflect a small proportion of the sun's energy back into space, counteracting the temperature rise caused by increased levels of greenhouse gases in the atmosphere which absorb energy and raise temperatures" (this can potentially be done by seeding the upper atmosphere with reflective particles, such as sulphide aerosols); and Carbon Dioxide Removal, which aims to "remove carbon dioxide from the atmosphere, directly countering the increased greenhouse effect and ocean acidification" (this can supposedly be achieved via ocean fertilization (e.g. algae blooms) or the spreading of "limestone, silicates, or calcium hydroxide in the ocean to increase its ability to store carbon" or through mechanical or organic means (e.g. biochar, turbines, etc.).[2]

A Brief History of Geoengineering borders on investigative journalism, anchored in the thesis proposed by Clive Hamilton's book *Earthmasters: The Dawn of the Age of Climate Engineering* and resulting from several months of researching the United States Patent and Trademark Office (USPTO), the

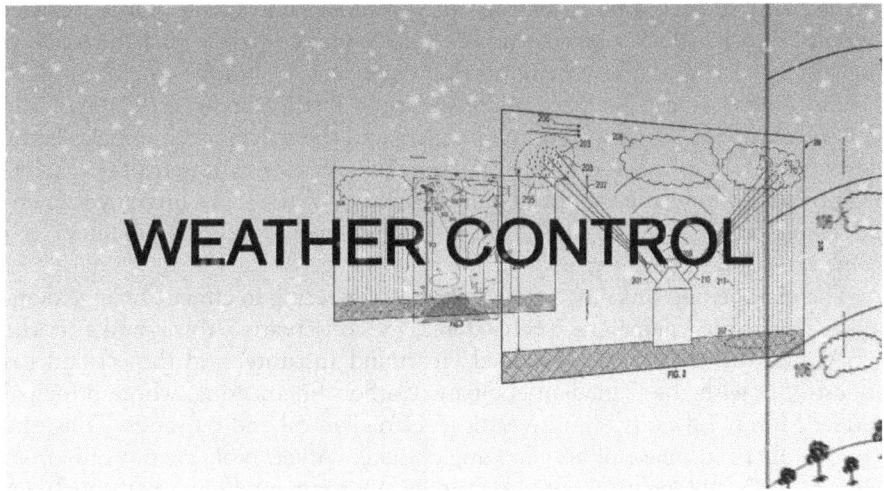

Figure 11.1 Video still from a *Brief History of Geoengineering*, 2015.
Source: © Mariana Silva and Pedro Neves Marques.

European Patent Office (EPO), as well as Google's patent global search engine, for applications on geoengineering technologies (Hamilton 2013). This research offered us the visual representations that ground the episode, in the form of patent file documents. Somewhere between an Apple commercial and an animated slideshow, patent application documents are shown in sequence, hovering towards the "camera", while the materiality of the xeroxed documents is rendered transparent against a background of digital CGI weather effects. CGI sunrays, snowstorms, thunder and heavy rain interrupt the succession of documents and follow the argument laid out by a voice-over. For instance, raindrops disturb the video's flatness, making it liquid; snow and direct sunlight block legibility. The soundscape adds another level of disruption that moves from bass-ridden pulses and beats to the foley-type sounds of punching of letters on a keyboard, thunderstorms and harsh winds. The effect was purposely inhuman – computerized and digital in an abstracted white, GCI space – playing with the aesthetics of infographic videos online.

Geoengineering purports to be a "technofix" for climate change mitigation that acknowledges the long-standing role of human intervention in weather systems. It has been proposed, not as a solution but as a complementary option alongside other measures, such as a cap on carbon emissions and large-scale recycling, to fight climate change. Given current predictions about the rise of global temperatures, proponents of geoengineering see it as unavoidable and urgent to buy us time in the fight against climate change. As such, it sits squarely within debates on the anthropogenic nature of the Anthropocene and the role of humanity in global weather systems.

Yet, rather than addressing the ontological questions underlying climate change and the recognition of the anthropogenic impact on global weather systems, the episode focuses instead on the economic interests and the marketing strategies behind climate engineering technologies. In following the trail of names, inventors and company addresses found in the patent applications, the episode outlines a who's who of the industry and the specific technologies being developed. This follow-the-money trail quickly reveals a tight network connecting private capital to academic laboratories and even the UN's Intergovernmental Panel on Climate Change, where the pro-geoengineering argument has gained traction in recent years.

The episode also links investment in geoengineering to climate change denialism marketing campaigns, orchestrated by conservative think-tanks in the USA, like the Exxon Mobil funded Heartland Institute, and the oil and gas industry, as with the Canadian company Carbon Engineering, whose principal funder Murray Edwards is an investor in Canadian oil and tar sands. Thus, the episode offers an image of an emerging climate market, profiting not only from climate change prevention but also from its environmental impacts through, for example, climate insurance services for regions affected by hurricanes or other environmental disasters.

In sidestepping the tremendous scale of climate change, which often tends to disempower people, in order to tackle its economics, the episode ties the Anthropocene to what historian Jason W. Moore has called "Capitalocene" (2015). While not necessarily denying a long geological view of the Anthropocene, which dates its beginning as far back as the Agricultural Revolution, in using the term "Capitalocene", Moore forces us to acknowledge the connection between our contemporary climate crisis and the rise of extractivist capitalism in the sixteen century and, later on, the Industrial Revolution.

Geoengineering, then, highlights a zone of conflict at the core of political debates around the Anthropocene. On the one hand, a techno-deterministic stance, claiming that if the Earth's systems are unnatural to start with, what is there to stop us adjusting the climate to our own necessities? On the opposite side, art historian T.J. Demos (2018) suggests, paraphrasing political activist Naomi Klein,

> [T]he lack of any regulatory protocol for climate interventions with transnational implications; its lock-in effect making it next to impossible to abandon the technology once it's been implemented; its anti-democratic basis in an era of globalism led by a handful of powerful developed nations; and, crucially, its directing of precious resources away from the *causes* of climate disruption, in favor of addressing *symptoms*.

This makes geoengineering questionable, to say the least. This economic view of the Anthropocene is precisely the take of *A Brief History of Geoengineering*. For, as stated in the episode, be it in the form of the climate itself or the capital invested in it, "weather knows no borders".

"Anthropocene issue": the breakdown of complexity

While researching for A *Brief History of Geoengineering*, we participated in the first Anthropocene Curriculum, a conference and campus of sorts bringing together 100 people purposefully spanning different fields of study and practice in the arts, humanities and science, held from 14 to 22 November 2014 at the transdisciplinary institution Haus der Kulturen der Welt (HKW) in Berlin, in order to subjectively document this vast gathering.

We won't go into detail about the organizational aspects, the objectives, and the assessment of this particular campus. Suffice it to say that the event was the result of HKW's two-year programme on the topic of the Anthropocene and the institution's involvement in the Anthropocene debate went as far as hosting a meeting of the Anthropocene Working Group, established by the International Commission on Stratigraphy to decide on the scientific dates and definition of the Anthropocene as a geological epoch.[3]

The broadness of knowledge at play on campus exemplified the complexity of the concept of the Anthropocene, how it entwines and confuses the roles and competencies of the humanities and the natural sciences, and how it breaks down the modern division between nature and culture. In a way, any attempt at documenting the totality of the campus would be as flawed as trying to encompass the scale of the Anthropocene discourse in a single video.

As such, we moved from room to room trying to follow at least two of the three daily workshops, from debates about the dating of the Anthropocene, which pitted geologists dating its origins to the Agricultural Revolution against historians who placed its beginnings at the rise of mercantile capitalism, to discussions about the reach and scope of computational climate models and the

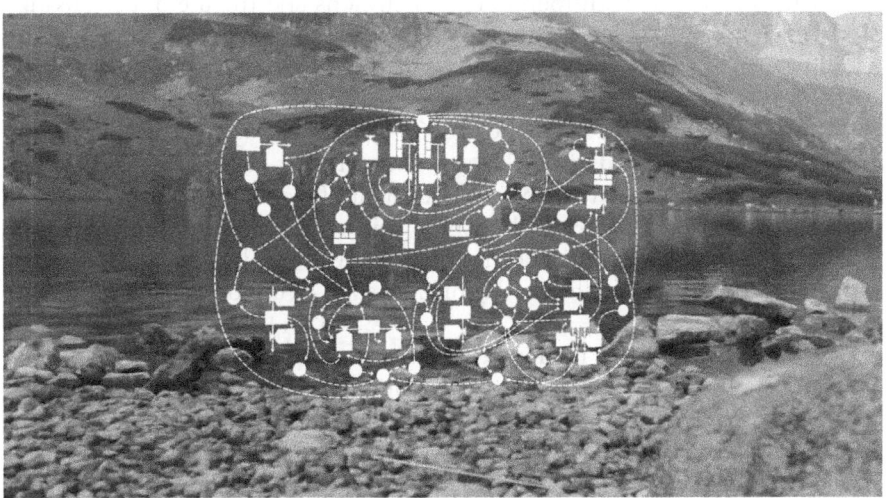

Figure 11.2 Video still from *Anthropocene Issue*, episode 1972, 2016.

Source: Artists' own images. © Mariana Silva and Pedro Neves Marques.

strong reactions this computational view instilled in the humanities. While doing so, we began to focus on the nuances between such exchanges, highlighting tensions between worldviews and methodologies between disciplines. We were filming details rather than wholes. This in-between space of both practices and knowledge became the core of the eight-episode series "Anthropocene Issue". It is perhaps unsurprising then that dialogue, be it in the workshops or in more private spaces, is the key mode of expression in the episodes.

Taking the cue offered by the workshop about the dating of the Anthropocene, we decided to title each episode in the series after a relevant date within this debate: the beginning of the Agricultural Revolution in around 10,000 BC; the *Limits to Growth* report, 1972; Chernobyl nuclear disaster, 1986; the launch of the first geostationary satellite into the Earth's orbit, 1964; the UN COP15 in Copenhagen, 2009; the seminal ratification of the Rights of Nature in the Ecuadorian Constitution, 2008; the Biosphere II experiment in the Arizona desert and the expected arrival of humans on Mars, respectively 1991 and 2027; and 2014 for the Anthropocene Curriculum itself. These dates reflected major topical points of contestation within the campus, as if each date expressed as much the desire as the incapacity of certain disciplinary fields (and practitioners) to suspend their disbelief – challenges that composed the Anthropocene. They also reveal the myriad ways one has to enter the topic of the Anthropocene, and the implications to the production of knowledge that each door poses.

"For an oil-free future": the role of fiction and other futures

While the *Anthropocene Issue* web series features our more documentary approach to date (e.g. with round-table discussions and the use of the talking-head format), it also includes experiments with fictional events and future scenarios in order to frame our present moment. This strategy was expanded in the 2016/2017 web series *Para um Futuro Livre de Petróleo* [For an Oil-free Future]. Shot in Portugal over the summer of 2016, the episodes both responded to and intervened in the heated public debate between the Portuguese government and local activist groups over the announcement of planned oil prospection and potential extraction onshore and offshore across the country – a debate that continues to this day, and which has largely been won by social movements, with the recent cancelling by the Portuguese state of up to 13 of the initial 15 planned drillings.

The oil extraction plans highlighted many contradictions between the government and civil society, between local and centralized power, and opposing visions of economic development at a time of increased tourism. Contracts with major oil companies (Repsol, Eni, Australis, Galp, Partex, Cosmos, and the controversial Portfuel) were issued without public consultation, and the opportunity to conduct a public debate around which energy plan the country should aim for was arrogantly sidestepped, despite a recent liberalization of the market in terms of electricity providers. As per an antiquated law dating from the beginning of the 1990s,

Figure 11.3 Video still from *Para um futuro livre de petróleo* (For an Oil-free Future), episode 1, 2017.

Source: Artists' own images. © Mariana Silva and Pedro Neves Marques.

environmental impact assessments were not obligatory ahead of signing the contracts. Furthermore, several activist and citizen-led movements (MALP, PALP, Futuro Limpo, Climáximo, Não ao Fracking Aljezur, Peniche Livre de Petróleo, among many others) pointed to the fact that oil and gas extraction would probably break Portugal's commitments at the level of the UN international climate agreement, namely the Paris accord. In other words, the movements said, while Portugal might arguably come to benefit energetically and economically from oil extraction, the decision to proceed with drillings at this stage of the world's history is ethically and economically questionable. For example, significant extraction only occurs after two decades of findings, when by then, however, and upon the UN's advice, a transition to renewables and other green sources of energy should already be in place.

The challenge we posed ourselves here was not so much one of representation, but rather of how to intervene in the political debate and be an active participant alongside social movements. Some of inhabitants' episodes are more informative or journalistic, but episodes such as those in *For an Oil-free Future* were designed to be circulated among local and/or international groups for their own use. Therefore, with *For an Oil-free Future* we began to collaborate directly with activists and sought to publish the episodes strategically, at poignant moments, so that they could be tools for action rather than a form of clickbait. As an example, we coordinated publishing the first episode a few days prior to an anti-extraction protest in Lisbon, adding to the momentum already in place.

At the narrative level, rather than focusing only on the current discussion, we decided to imagine its consequences in the future. In doing so, we hoped to tie a little remembered, and recorded, history of environmental activism in Portugal to current and future struggles while also raising the question of which and whose future is at stake. For this reason, we decided to place the narrative of the episodes 20 years into the future, depicting a dystopian scenario, visually characterized by an eerie and atmospheric reddish hue, where extraction has indeed occurred and the global consequences of climate change are felt along the Portuguese coast. The first episode used archival footage to tell a brief history of oil spills in Portuguese waters and environmentalist responses to the spills, which was then intercut with scenes shot at the national oceanarium, so as to simultaneously recall but also signal the hypocrisy of a national imaginary tying the country to the oceans. This was also a way to address environmentalist concerns with the risk of future oil-related disasters, while also acknowledging that such disasters had already occurred in the past. The remaining episodes were based on interviews with experts, which were then re-enacted with actors and staged as if taking place in the 2030s, and were structured around Rob Nixon's notion of slow violence; that is, that alongside visible, direct and immediate violence (in this case the possibility of future oil spills, for example) one must also consider more invisible and longer term modes of violence, in this case in the form of pollution and toxicity resulting from the petrochemical industry and its impacts on human and non-human health (Nixon 2013). Lastly, the fifth and final episode in the series was shaped as an exposition of some of the principles behind the campaign Jobs for the Climate, which pressures the state to transition workers related to the petrochemical and other polluting industries to jobs in the renewables sector, thus building bridges between environmentalist concerns and workers' unions.[4]

"What is deep-sea mining?": giving image to what is to come

Our most recent web series on environmentalist issues began with an invitation from Margarida Mendes, contemporary art curator and fellow collaborator of inhabitants, and her activist work against deep-sea mining (DSM). Together with different biologists and the support of already existing NGOs (such as Seas at Risk and SCIAENA), the first Portuguese environmentalist anti-DSM movement *Oceano Livre* [Free Ocean] was founded in 2017, of which Mendes is a founding member.

But what is DSM? It is:

> A new frontier of resource extraction at the bottom of the ocean, set to begin in the next few years. Deep sea mining will occur mainly in areas rich in polymetallic nodules, in seamounts, and in hydrothermal vents. Mining companies are already leasing areas in national and international waters in order to extract minerals and metals such as manganese, cobalt, gold, copper, iron, and other rare earth elements from the seabed.

Main sites targeted for future exploration are the Mid-Atlantic Ridge and the Clarion Clipperton Zone (Pacific Ocean) in international waters, as well as the islands of Papua New Guinea, Fiji, Tonga, New Zealand, Japan, and the Portuguese Azores archipelago. Yet, potential impacts on deep sea ecosystems are yet to be assessed by the scientific community, and local communities are not being consulted. The prospect of this new, experimental form of mining re-actualizes a colonial, frontier mentality, redefining extractivist economies for the twenty-first century.[5]

This five-part web series ranges from more explanatory episodes to regionally specific episodes, allowing our audience a step-by-step panorama of DSM, what it is and where it is set to happen in the near future. As of the moment, two out of five episodes have been published. Here again we were faced with a question of scale, given that DSM is being planned for both the Atlantic and Pacific oceans and any reporting on the subject demanded of us a dialogue with either scientists or anti-DSM activists worldwide. Because of this, we soon found ourselves in a myriad of Skype calls and email exchanges connecting Lisbon to New York, Boston and Arizona to Brussels, Papua New Guinea and Australia, consulting with anti-deep-sea mining campaigns already on the ground, marine biologists, geographers, anthropologists, and lawyers with expertise in ocean law, as well as with funding partners, such as the art institution TBA21 – Academy.[6]

Figure 11.4 Video still from *What is Deep Sea Mining? Episode 1: Tools for Ocean Literacy,* 2018.

Source: Artists' own images. © Mariana Silva and Pedro Neves Marques.

Perhaps even more challenging is the fact that, despite the increasing interest of many nation-states and both state and private financial investment companies in the technology, DSM is yet to occur anywhere on the planet. In the words of *Oceano Livre*:

> Potential for mineral wealth in the deep ocean was identified as early as the 1960s. However, attempts to mine have been unsuccessful so far, as the reserves have proved too expensive to get to and difficult to mine. It is only very recently, as technological advancement has been matched by escalating commodity prices and demand, that the highly speculative practice has begun to be considered economically viable by some companies, and territorially important by some countries.[7]

As of the moment, DSM is simply a new, imagined future extractivist horizon. This does not mean that its consequences are not already being felt, as, for example, in Papua New Guinea, where the local government together with mining company Nautilus are yet again sidelining and disempowering indigenous communities and environmentalists on the case of the deep-sea Solwara 1 mining field. But it does pose a challenge not only in what concerns information but also and more importantly to mobilizing people against it. In other words, how to report and create images of contestation against an industry that does not yet exist?

In a way the web series *What Is Deep Sea Mining?* poses a similar challenge to *For an Oil-free Future*. With deep-sea mining however, the matter is even more abstract, as, while the environmental consequences of DSM are yet unknown, biological knowledge of the deep seabed is also scarce. Examples of oil and gas extraction imagery abound, as well as of accidents such as oil spills or consequences such as sea-level rise. But given that DSM is still in its experimental phase, all one has are predictions, a small set of deep-sea experiments, and comparisons to dredging used in industrial fishery. On the other hand, efforts to map the oceans have a long history, but knowledge of the deep sea is fairly new. While it was only with the advent of sonar during World War II that the first tridimensional geological maps of the seabed began to take shape, target areas for DSM, such as hydrothermal vents (the main object of DSM in the Azores), manganese nodules fields and cobalt-rich ferromanganese crusts found on seamounts (mainly in the Pacific), only began to be studied from the late 1970s on. Whereas certain regions of the sea floor were until recently thought to be mostly lifeless, some (such as hydrothermal vents) are actually rich in life forms, some of which are endemic and regarded by marine biologists as seminal for understanding the genetic origins of life on Earth. This makes any image-making effort speculative.

This challenge to image-making, alongside the exchange of knowledge between different anti-DSM struggles across the world's oceans – what we, knowing that targeted territories are by and large islands, have termed "island solidarity" – is the key guideline behind *What Is Deep Sea Mining?*.

Thus, for episode 1 in the series entitled *Tools for Ocean Literacy*, we went back to an old time image-making favourite film of ours, 1977's *Powers of Ten* by Charles and Ray Eames, wherein a camera pans out from the human scale to a galactic scale and back to the molecularity of atoms. From it we took the challenge of imaging scales and the single camera pan as a strategy to cross, in this case, the invisibility of human-imagined frontiers that compose the different orbital strata (from sea level to the outer orbit: the troposphere, stratosphere, thermosphere, low and high Earth orbits) as well as the different ocean regions (the epipelagic, mesopelagic, bathypelagic, abyssopelagic and hadopelagic zones). Meanwhile, two voice-overs narrate a brief history and introduction to what is DSM, alerting the audience to its potential threats and pointing to the activism of social movements.

For episode 2, entitled *Deep Frontiers*, we invited anthropologist Stefan Helmreich, who has done extensive ethnographic work among marine biologists, to write its script. The episode focuses on deep-sea life, resorting to a ROV (remotely operated vehicle) submarine aesthetics and its mechanical point of view as it surveys the pitch-black bottom of the sea: bioluminescent silhouettes of deep-sea creatures fade in and out of the darkness, interrupted by a blinding flash of light, which strikes the "flat image" (that is, the screen). Helmreich, author of the book *Alien Ocean: Anthropological Voyages in Microbial Seas*, asks us to familiarize ourselves with the alienness of life forms found at the bottom of the oceans, from tubeworms to bacteria, in an attempt to widen the scope of our human empathy beyond mammals and animals anatomically closer to us.

The question of image-making is, of course, intimately related to political struggles. Images both shape and are shaped by the visual cultures that surround them. In the case of deep-sea mining this does not mean simply the future one wants (or inversely a future one wishes to avoid). A major drive behind the rush for the seabed is the demand for minerals used in renewable technologies, as well as those used in our current digital revolution, from computer processors to cellphone batteries and screens. The question of an image of the future in the context of DSM is therefore infrastructural, and proves a challenging dilemma. To struggle against deep-sea mining demands an inquiry into the sustainability of our technological future, paradoxically enough a future where "green" technology has been fully implemented.

Conclusion: images in struggle

Our objective at inhabitants has been to combine investigative reporting on given political issues while also rupturing with expectations about how the form of such informative videos should look like. To us this is a challenge in connecting journalism with the visual arts. Unlike other non-professional online platforms, we are not so much interested in a coherent theme as in maintaining rigor and formal experimentation in our approach, while also opening up to a network of agents – activists, academics, filmmakers – already working on the ground, bringing them into social causes. These dialogues have also taught us that, in terms of distribution, instead of aiming for a video to go "viral" it is rather more interesting to be in touch with activist groups and to make sure they get access to the content, that they find it useful, and that we time each episode launch according to their calendar of events. This puts us in a position of asking us to both believe and question the techno-utopianism underlying the social media hype, the market and algorithmic logic of virality, and the recent economic investment by news outlets in video content.

Looking back at the above episodes, however, it becomes clear to us that, despite its differences, all in all we have been dealing with visibility and challenges to image-making. From the abstraction of geoengineering and the planetary and geological scale of the Anthropocene to the future of oil extraction and deep-sea mining, the challenge has been throughout to give an image to such topics, and in many cases the future. This is, in fact, a double challenge. On the one hand, how to act journalistically; that is, how to detail information and hopefully clarify nebulous topics in a way that is clear and concise for our audience. On the other hand, how to work with images to empower that same audience to imagine other possible worlds, to look for solutions and join political struggles. As recent developments in the world of social media have taught us, the making of images and narratives online has the potential for change and social influence. This is as much a question of content as of form, and as such makes any encounter between contemporary art and journalism a site of politics.

Notes

1 http://inhabitants-tv.org/about.html.
2 See www.geoengineering.ox.ac.uk/what-is-geoengineering/what-is-geoengineering/.
3 For more information see the project's website: www.hkw.de/en/programm/projekte/2014/anthropozaenprojekt_ein_bericht/anthropozaenprojekt_ein_bericht.php.
4 For Jobs for the Climate see www.empregos-clima.pt/.
5 Inhabitants website: http://inhabitants-tv.org/feb2018_whatisdeepseamining_ep1.html.
6 Among the NGOs and groups we contacted are: Seas at Risk, Deep Sea Conservation Coalition, Deep Sea Mining Campaign, Deep Sea Mining Watch and Solwara Warriors.
7 See the website of Oceano Livre: http://oceanolivre.org/en/o-que-e-a-mineracao-em-mar-profundo.

Bibliography

Baudrillard, Jean. 1991. *The Gulf War Did Not Take Place*, originally published as *La Guerre du Golfe n'a pas eu lieu*. Paris: Éditions Galilée.
Demos, T.J. 2018. "To Save a World: Geoengineering, Conflictual Futurisms, and the Unthinkable". *e-flux journal* 94. www.e-flux.com/journal/94/221148/to-save-a-world-geoengineering-conflictual-futurisms-and-the-unthinkable/.
Elsaesser, Thomas. 2004. *Harun Farocki: Working the Sight-lines*. Amsterdam: Amsterdam University Press.
Hamilton, Clive. 2013. *Earthmasters: The Dawn of the Age of Climate Engineering*. New Haven, CT, and London: Yale University Press.

Latour, Bruno. 1993. *We Have Never Been Modern*. Cambridge, MA: Harvard University Press. Originally published in French in 1991.

Moore, Jason W. 2015. *Capitalism in the Web of Life*. New York: Verso.

Morton, Timothy. 2013. *Hyperobjects: Philosophy and Ecology after the End of the World*. Minneapolis: University of Minnesota Press.

Nixon, Rob. 2013. *Slow Violence and the Environmentalism of the Poor*. Cambridge, MA: Harvard University Press.

Panofsky, Erwin. 1991. *Perspective as Symbolic Form*. New York: Zone Books. Originally published in German in 1927.

Serres, Michael. 2007. *The Parasite*. Minneapolis: University of Minnesota Press.

Taussig, Michael. 1993. *Mimesis and Alterity: A Particular History of the Senses*. New York: Routledge.

12 Troubled gardens

Nature–technoculture binary and the search for a Safe Operating Space in Hayao Miyazaki's *Mononoke Hime*

Ivo Louro and Ana Matilde Sousa

Introduction

Gardens featured in television and films are more than just opportunities for showcasing little patches of green on the screen. They reflect the beliefs of characters and provide valuable clues about in-world meanings, emotions and associations, much like they do in real life. Now, the representation of gardens in fiction takes on a renewed importance as we enter the Anthropocene, a (proposed) new epoch in which anthropogenic impacts on the Earth's ecosystems are transformed into planetary forces, capable of affecting climate change and other global phenomena. Gardens hold a great potential for the exploration of anthropocenic poetics, working as miniature universes that encapsulate human relationship(s) to nature, technology and itself. This is especially true in animation, where, unlike live-action films, visions of "nature" are constructed from scratch through drawings, sound design and the illusion of movement. This chapter focuses on one such case to explore the representation of gardens: *Mononoke Hime* (もののけ姫), or *Princess Mononoke*, Studio Ghibli's 1997 box-office hit written and directed by the guru of Japanese animation, Hayao Miyazaki. The choice may, at first glance, seem odd. For one thing, *Mononoke Hime* is set in the historical past, albeit a fantastic one. It does not present us with visions of futuristic technology, nor does it fit well within the genre of eco-disaster cinema. For another, a cursory look at *Mononoke Hime* will not return any prominent or even recognizable gardens. Nevertheless, we hope to demonstrate that (1) the centres of human and natural spaces in *Mononoke Hime* are underlined by garden affects and aesthetics, as enclosed spaces of otherness whose meaning emerges from their connectivity; (2) the film's message and resolution rests upon the negotiation of "basic forces, forms and ideas of man and nature, control and chaos, stasis and growth, the wild and the civilized, escape and involvement, ideals and reality" (Helphand 1997, 101) found in garden theory. *Mononoke Hime* echoes this interdependency of natural and human agencies, but evades ecomodernist fantasies of the Earth as a well-administered technogarden,[1] with humans serving as global gardeners in a good Anthropocene. Instead, the concept of Safe Operating Space (Steffen *et al.* 2015) captures more adequately the complex entanglement of nature, culture, gender and race in *Mononoke Hime*, which sets it apart from other environmental films (even within Miyazaki's oeuvre), endowing its gardens with particular weight in terms of anthropocenic poetics.

As anime and manga critic Susan Napier (2005, 248) suggests in her analysis of *Mononoke Hime*, the film reflects "the extraordinary array of pluralities that suggest the ever more complex world of the twenty-first century". The Anthropocene as a "genre" of speculative fiction often puts on the shoes of sci-fi, justifying Ursula Le Guin's popular oxymoron that "science fiction is not predictive; it is descriptive" (i.e. that it is about the present, not the future). On the contrary, *Mononoke Hime* travels back in time to discuss our current state of affairs, namely to the Muromachi period (fourteenth to sixteenth centuries) of Japanese history marked by continuous warfare as much as by significant material growth in terms of technology, economy and urbanization. Nature, in *Mononoke Hime*, preserves the markers of a Romantic pastoral framework – pristine, ultimately redemptive, if ruthless and awe-inspiring, opposed to the metropolis. However, a deeper inquiry into the film's characters and locations makes it clear that the nature–technoculture binary in *Mononoke Hime* is also destabilized, troubled, by averting the stultifying formulas in which many environmental films indulge; for instance, the representation of industrialization as an autonomous evil, driving the destruction of nature for profit or out of pure villainy. In addition, unlike environmental films such as *Silent Running* (1972), *Who Framed Roger Rabbit* (1988), *Ferngully: The Last Rainforest* (1992) or *Avatar* (2009), *Mononoke Hime* does not offer a neatly packed resolution. It recognizes that the complex ethical quandaries of coexistence on Earth cannot be settled by adopting black-and-white solutions that flatten the wrinkles of human and non-human relations. This chapter will focus on two locations in *Mononoke Hime*, the two "gardens" epitomizing the opposite sides of the film's central conflict: Eboshi's secret garden in the city of Tatara, and the pool of the *shishi-gami* ("elk god") in the heart of the forest. These sites are contrasted, but also paralleled, different, but in many ways similar; but above all they are co-constitutive. Understanding their entanglement is key to grasping the film's socio-ecological message.

The chapter is divided into three sections. After we introduce the events at the beginning of the film we focus on Eboshi's garden as a space of otherness, and its connection to a broader cosmogony of humans and the natural world. After that, we analyse the pool of the *shishi-gami*, focusing on its relationship to Japanese gardens. Finally, we address the concept of Safe Operating Space, originating from Earth Systems Sciences, and its relevance for understanding the resolution of *Mononoke Hime* within the broader Anthropocene debate.

Curse and journey to the West

Mononoke Hime follows a boy called Ashitaka in his quest to lift a cursed mark on his arm. Ashitaka received the curse while defending his village from a wrathful *tatari-gami* (タタリ神), or cursed god, a monster covered in black tendrils, that the boy manages to shoot and kill. As the tendrils melt away, however, they reveal a mortally wounded boar god (イノシシ神, *inoshishi-gami*) underneath. The boar god dies cursing humanity, and from inside its body the villagers retrieve the apparent source of its agony: an iron bullet. When the village oracle reveals that Ashitaka's cursed mark will spread and claim his life if left untreated, he has no choice but to follow the boar's trail to the West in the hope of finding a cure. In a beautiful sequence where Miyazaki showcases his passion for evergreen landscapes, Ashitaka rides through valleys, mountain passes and other villages. Eventually, the boy realizes that the curse has not only endowed him with superhuman strength, but that his arm possesses a violent will of its own, as Ashitaka defeats two samurai bandits who attack him on the way with unexpected ease and brutality.

At last, Ashitaka arrives in a land where two factions wage war. On one side, the denizens of a primeval forest protected by the *shishi-gami* (シシ神), or Forest Spirit, an elk god presiding over the life cycles of animals and plants. On the other, Tatara, a fortress home to an ironworks-cum-arms factory led by the fierce Eboshi and home to a community of social outcasts. To secure their city from external attacks, Eboshi starts a deforestation campaign, triggering the hatred of the forest gods, who in turn seek to destroy her and Tatara. Later, we learn that the Emperor of Japan has blackmailed Eboshi into bringing him the *shishi-gami's* head, which he believes will grant him immortality. Among the denizens of the forest, the 300-year-old wolf goddess Moro, and San, a human girl raised by the wolves, are particularly invested against Eboshi. Despite being human, San hates mankind, even calling herself a wolf. For this reason, the villagers nicknamed her Princess Mononoke, referencing the vengeful *mononoke* (物の怪) spirits common in Japanese folktales, which were believed to possess people and make them suffer. After rescuing two soldiers injured in a battle against Moro, Ashitaka is welcomed by Eboshi as a guest, but not before sighting Moro and Princess Mononoke from a distance. When she first appears, San is dressed in tribal attire, with her mouth covered in blood as she sucks the poison from Eboshi's bullets out of Moro's neck – who she calls "mother" and by whom she is called "daughter". In this way, *Mononoke Hime* endorses a long-standing ideal of femininity as primitive and closer to nature, fascinating and even wild or threatening (San behaves animalis-tically on various occasions and her appearance is marked with tribal signifiers, including a clay mask, a cape made of wolf's fur, red war-paint on her face and a fang necklace). The film's imbroglio begins when technoculture, instead of being represented as a macho military-industrial complex or an evil witch/monster (e.g. *Avatar* or *Ferngully*), is also gendered feminine, maternal and othered like nature. This choice, embodied by Eboshi, is fully exposed in the sequence where she takes Ashitaka to her secret garden in Tatara.

Eboshi's secret garden

In Tatara, Eboshi asks Ashitaka about the purpose of his journey. Ashitaka shows her the iron bullet and his cursed arm, telling her that she should recognize the weapon that transformed the boar god into a *tatari-gami*. When Eboshi asks Ashitaka what he will do if she tells him the truth, the boy replies that he will "ascertain everything with unclouded eyes, and decide".[2] Ashitaka's idealism elicits a hearty laugh from Eboshi, who agrees to reveal her secret despite her bodyguard's protests. Eboshi then takes Ashitaka around Tatara, passing workshops where workers go about their daily business. Their activities are rendered in clean, glassy sounds rather than the harsh noises commonly associated with the manufacturing industry. Ashitaka stops briefly at the gates of a huge kiln operated by – we learn in a previous scene – liberated ex-prostitutes, who have chosen to follow Eboshi after she bought their brothel contract and set them free. Finally, when they arrive at a secluded corner of Tatara where the sick

Figure 12.1 Eboshi and her garden in Tatara. Video still from Hayao Miyazaki's *Mononoke Hime* directed by Hayao Miyazaki and produced at Studio Ghibli, Inc.

Source: © 1997 Nibariki TNDG.

ward is located, Eboshi announces: "This is my garden, that everyone fears and dares not approach. If you wish to know the secret, come."³ Within the precincts, there is a modest vegetable garden (the only time we see greenery in the city) and a hut, inside which Ashitaka finds a solemn work scene where bandaged lepers are developing a new model of light rifles.⁴ The interaction between Eboshi and the lepers is courteous and, at times, endearing. Eboshi apologizes for interrupting and making them hurry, promising to have *sake* delivered to them later on. In turn, the lepers proudly show off their creations to Eboshi, joking about her being "scary" (*Kowaya, kowaya!*) and wanting to destroy the country. Eboshi also states that the rifles are to be used by the women of Tatara to crush both monsters and samurai. The rifles play a central role in the community's self-determination and defence against both the neighbouring warlords who lust after Tatara's resources, and the onslaughts of nature.

Despite all this, Ashitaka becomes enraged. He blames Eboshi not only for destroying the woods and turning the boar god into a monster, but also for perpetuating the violence by manufacturing more guns. The boy's cursed arm pounces with killing intent. He forcefully stops himself from attacking, aware that more bloodshed would neither lift his curse nor solve the conflict. At that moment, a bedridden leper known as *osa* ("chief") begs Ashitaka not to kill Eboshi, saying that although he understands Ashitaka's rage, Eboshi is the only one who treats them, the lepers, humanely:

> Young man, I am also cursed, so I understand your anger and sorrow all too well. I understand, but I beg you, please do not kill her. She is the only person who treated us like human beings. Without fear of our disease, she washed my rotting flesh, she bandaged me.⁵

Osa adds: "To live is extremely difficult and painful! I curse the world, I curse the people, but still I want to live!"⁶ Such affirmations of life are a constant throughout the film, with both sides of the conflict repeating the idea that, as long as they are alive, they can somehow recover and live on. In fact, the film's promotional tagline was "*Ikiro*" (生きろ), meaning "Live" or "Survive". Here, the Japanese verb "to live" or "to exist" (生きる) appears in the seldom-used command form, conveying an urgent imperative or encouragement to keep on living.

By drawing a comparison between Ashitaka's curse and his own disease, *osa*'s words reframe the people of Tatara as cogs within a broader cosmic continuum. *Mononoke Hime* equates Tatara's struggle to survive in adverse conditions (including the production of firearms) with that of the forest's animals and gods, challenging us to consider the issue of survival and habitability as a complex system evading clear-cut solutions. As Napier (2005, 246) points out,

> Miyazaki problematizes the issue even further by making Tatara not just a site of industrial production but a site of weapons manufacturing. [...] However, it is these weapons that give employment to Tatara's outcast citizenry, who surely have as much right to survival as the denizens of the forest.

Figure 12.2 Osa speaking to Ashitaka. Video still from Hayao Miyazaki's *Mononoke Hime* directed by Hayao Miyazaki and produced at Studio Ghibli, Inc.

Source: © 1997 Nibariki TNDG.

In this sense, military technology encapsulates the destructive and creative drives embedded in many industrialization processes, through which people both improve or degrade their and others' living conditions.

Visually, Eboshi's garden offers none of the compelling elements that, more often than not, are present in the representation of gardens in film: the colourful flowers, the lush greenery, the natural light. Rather, Ashitaka's visit takes place at night, imbuing the rows of vegetables and small trees with a restrained, somber glow. Nevertheless, Eboshi's garden retains the crucial element of enclosure in garden aesthetics. As Kenneth Helphand puts it,

> In virtually all tongues the etymology of garden describes an enclosed piece of land. This demarcation and separation is a recurring theme: sense of enclosure – the garden as a place apart – a different place, pervades all of our garden thinking.
>
> (Helphand 1997, 101)

This idea of the "garden as a place apart" is palpable in the gated precinct of Eboshi's garden, encircled by high fences separating it from the outside. Eboshi's garden is a place of otherness, a non-hegemonic "safe space" outside sexist and ableist privilege, where marginalized groups (prostitutes, lepers) are nurtured and empowered according to the principle of "from each according to his ability, to each according to his needs". The fact that Eboshi's garden is a secret "apart" from the rest of Tatara allows for its revelatory impact on the film, even if the garden itself appears for only a few seconds on screen. From that point onwards, Eboshi, who at first appears to be the ruthless face of industrial progress and warmongering, is refashioned as a Marxist feminist who opposes the inequality derived from capitalism (the warlords) and patriarchy (the Emperor). More than an embellishment or visual spectacle, Eboshi's garden encapsulates the essence of Tatara as an arena of development, both technological and societal – as well as the need to dig deep into the heart of matters if we are to see with "unclouded eyes", reflecting Miyazaki's own shifting perception of the world at the time, shaped by events such as the Gulf War and the Yugoslav Wars.[7]

The framing of Eboshi's garden between the prostitute's kiln and the lepers' shack speaks of the way in which Miyazaki chooses to convey his message in *Mononoke Hime*. He emphasizes the dynamic relationships among characters, places and events, more than their existence as separate things. This resonates with concepts from Shinto and Buddhist mysticism, for which Miyazaki is renowned for integrating into a loose "Japanese folk religiosity" (Mazur 2011, 324). For instance, *osa*'s "naturalization" of Tatara's struggle evokes the cyclical-ity of Saṃsāra (the eternal wheel of rebirth in which all life and matter parti-cipates) as much as techno-animist notions derived from Shinto-infused cosmopolitics, blurring the boundaries between human, animal, spiritual and mechanical. Whereas Western notions of technology and science are radically separated from pre-modern "spiritually-informed ideas of worldly co-habitation", techno-animism wields both fields together, fostering hybridism between human and non-human, the worldly and the otherworldly (Jensen and Blok 2013). At first glance, Eboshi's garden does appear to be a space oriented towards produc-tion (of vegetable and guns), devoid of mystical flares. But its form and content are embedded within techno-animism, in a broad sense, less as the attribution of "aliveness" to technological objects, than the insertion of the material uni-verse – even that resulting from industrial practices – into "an integrated natural world that includes the human" (Gifford 1999, 148). This integrated worldview is crucial for understanding Eboshi's Tatara, especially in relation to our next "garden": the pool of the *shishi-gami* at the heart of the forest.

The pool of the *shishi-gami*

If Eboshi's garden represents the heart of the factory-city, the place where the *shishi-gami* dwells deep in the woods is the "center of the film's fantasy space [...] that stands in uncanny opposition to the civilization of Tatara" (Napier 2005, 242). While Tatara stands for technoculture, the woods that Moro, San and the other animals and forest spirits inhabit are rendered as the cultural nature of fables, where anthropomorphized animals talk, produce artefacts and engage in tribal modes of organization. Miyazaki draws both from Shinto beliefs and the writings of ethnobotanist Sasuke Nakao – who argued that Japan was once covered in lush vegetation before the advent of the rice culture that now domi-nates the country's landscape – to conceive of the glossy-leaved forest as a "buried archetypal memory" that resonates deeply within the hearts and minds of Japanese people (Napier 2005, 242–243). Moreover, the natural world in *Mononoke Hime* does not stand in monolithic harmony, nor is it united against the human invaders, as it does in films like *Avatar* (in which the indigenous Na'Vi fight united against an evil mega-corporation). On several occasions we are shown different factions at odds, that disagree with and even attack one another: the wolves, who decide to help Ashitaka after the *shishi-gami* cures his wounds; the boars, whose *bushido* pride drives them to a final suicidal attack against Tatara's army; or the apes that blame San for bringing ruin to the forest by being neither animal nor human. The *shishi-gami*, however, seems to inhabit a different plane altogether, a pre-cultural nature differentiated from the rest of the forest by its looks, its lack of human language and the place where it resides: a clearing in the forest with an island and a pool of tranquil water.

The archetypal forest encloses the pool of the *shishi-gami*, but the latter is formally differentiated from the former. Like Eboshi's garden, which functions as Tatara's secret core, the *shishi-gami*'s pool is also "a place apart", a discontinuity in the landscape, resembling a Japanese garden like the one at Saihō-ji temple in Kyoto. Instead of a vegetable garden like Eboshi's, the *shishi-gami*'s is a moss garden with a central pond and a small island, whose luminous *sugi-goke* moss casts a diffuse glow across the woods. The palette is minimal, restricted to deep grey-greens, and the light is dappled, bathing the entire pool in shadows. According to Marc Treib (1989, 101), the traditional Japanese aesthetic of "*yūgen* derives from this mental conflation of moss, water, shade and reflection". The resulting sense of mystery, profundity and grace matches the *shishi-gami* itself, a chimerical collage of animal parts with a sphinx-like smile that does not

Figure 12.3 View of the *shishi-gami*'s pool. Video still from Hayao Miyazaki's *Mononoke Hime* directed by Hayao Miyazaki and produced at Studio Ghibli, Inc.

Figure 12.4 Moss gardens of the Saiho-ji temple in Kyoto.
Source: © 2004 Ivanoff. Source: https://commons.wikimedia.org/wiki/File:Saihouji-kokedera01.jpg

speak and whose appearance is often accompanied by silence. In this sense, the archetypal forest around the *shishi-gami's* pool functions as *shakkei* or "borrowed scenery" – a Japanese gardening practice of Chinese origin "where an outside view is incorporated to become part of a garden scenery" (Kuitert 2015, 32), seeking a continuity between interior/exterior spaces. As argued by Wybe Kuitert (2015, 33), the original sense of the Chinese term[8] refers to a process of mutual exchange in which "borrowing from the landscape goes together with the landscape that lends". This is important in that the *shishi-gami's* pool and Eboshi's garden are "places apart" that exist in interaction with their surroundings (the forest and the city, respectively) and with each other; and whose revelation alters the audience's aesthetic and affective experience of the world of *Mononoke Hime*.

Miyazaki highlights this garden reciprocity by establishing various parallels between Eboshi's garden and the *shishi-gami's* pool. On a formal level, both have an arena-like design encircled by clear boundaries: Tatara's fences in Eboshi's garden, and trees, shrubs and massive boulders in the pool of the *shishi-gami*. Moreover, because Ashitaka visits Eboshi's garden during the night, both are bathed in silence and a shadowy light. Like the *shishi-gami's* pool, Eboshi's garden is contrasted with the hustle and bustle of Tatara, namely the sounds from the workshops and the kilns that Ashitaka hears on the way to the sick ward, giving the impression of a refuge or sanctuary. Symbolically, Eboshi's garden and the *shishi-gami's* pool are framed as "creative-destructive" sites of life and death, nurturing and suffering. Both are spaces of otherness, embodying a maternal (if androgynous) refuge, while existing in specific social-cultural conditions. In Eboshi's garden, the sick ward where the lepers are cared for and protected – the very site of Eboshi's *caritas* – doubles as a research and development unit of gunsmiths. In turn, the *shishi-gami* is an ambivalent creature, harbouring an underlying violence despite its gracefulness. According to Miyazaki's storyboard notes, its "expression looks both benevolent and cruel" (Miyazaki 2014b, 101). This ambivalence is encapsulated in the memorable sequence in which the *shishi-gami* saves Ashitaka's life after the boy is mortally shot in the process of saving San from Tatara's people. In return, San places Ashitaka on the shore of the *shishi-gami's* island, sticking a branch into the ground like a grave marker, and waits. When the *shishi-gami* appears, the plants sprout around its hooves, and grow and wither with each step. The *shishi-gami* sucks the life out of the branch and breathes it into Ashitaka's body, sparing the boy's life without lifting his curse. Although the sequence is dreamlike and peaceful, it hints at the violent, unfathomable, indifferent side of nature. The *shishi-gami's* designs as giver and taker of life cannot be grasped by those outside its plane of existence.

In addition, by portraying the "center of the film's fantasy space" as a Japanese garden, Miyazaki recognizes the *shishi-gami's* pool as a filmic construction resulting from the director and his team's landscape gardening. As critics like Thomas Lamarre (2009, 104–105) point out, *Mononoke Hime*'s background landscapes mobilize detail, composition, colour and luminosity to convey nature through a painterly illusion of depth, contrasted with the flatness of characters. Like Japanese gardens, these landscapes seek to "underplay or even conceal all signs of human contrivance" (Treib 1998, 71) to emulate the complexity of an uncontrollable nature. The fact that the centre of Miyazaki's nature is a "garden", however, amounts to more than a view that nature does not exist beyond human constructs. Rather, the opposite is at stake. The Baradian concept of *intra-action* can help us reconcile the garden as a place apart, gated or otherwise separated from its environment, with the "dynamism of Nature (rather than inert plenitude) that provides an absolute and abiding frame of reference" (Lamarre 2009, 105) in Miyazaki's films. For Karen Barad (2007, 33), the term *intra-action* "signifies the mutual constitution of entangled agencies. That is [...] the notion of *intra-action* recognizes that distinct agencies do not precede, but rather emerge through, their intra-action." Eboshi's garden and the *shishi-gami's* pool are apparently distinct, even dichotomous (city vs. nature), but any true understanding of them necessarily entails a comprehension of the ways in which they "are only distinct in relation to their mutual entanglements" (Barad 2007, 33). They are, so to speak, a "becoming-garden" or mutually constitutive gardens where one is not what the other lacks. If, at the beginning of the film, Miyazaki frames Tatara as a "malignant tumor" (Miyazaki 2014b, 62), poisoning the pristine environment that surrounds it, this vision of monolithic evil is systematically deconstructed from the moment that Ashitaka sets foot inside the city – with Eboshi's garden sequence as a major turning point. In the world of *Mononoke Hime*, humans, animals and gods become a part of the same cosmopolitical fabric, as *osa*'s words hint at. It is not that the distinction between nature and technoculture ceases to apply, but its terms are destabilized, challenging the audience to pursue alternative schemes to accommodate a world no longer viewed in black and white.

The meaning of Eboshi's garden and the *shishi-gami*'s pool emerges in relation one to the other, calling into question clear-cut distinctions of "good" and "evil". This does not mean that the film is neutral. Indeed, despite his sympathetic depiction of Tatara's people, Miyazaki uses dark purplish-red and red-brown colours to establish a continuity between Tatara and Ashitaka's curse. Nevertheless, although Miyazaki draws heavily from the legacy of pastoralism, it feels inadequate to characterize *Mononoke Hime* as a pastoral work, as is often the case. Rather, *Mononoke Hime* raises questions distinctive of what literary critic Terry Gifford calls the "post-pastoral" mode. Drawing from Leo Marx's "complex pastoral", Gifford (2012) conceives of post-pastoralism as an attempt to overcome the opposition between the pastoral and anti-pastoral modes – neither romanticizing nature as redemptive idyll nor cultivating a cynical vision of the natural world. Post-pastoralism advocates that conflicting forces can and should be articulated through a non-binary framework, taking into consideration the complex, often contradictory aspects of humanity's relationship to nature, as well as "nature" itself as a problematic category. According to Gifford (2012, 21–27), post-pastoral works typically raise the following questions among audiences: (1) Can awe in the face of nature (e.g. landscapes) lead to humility in our species, reducing our hubris? (2) What are the implications of recognizing that we are part of nature's creative-destructive processes? (3) If our inner nature echoes outer nature, how can the latter help us understand the former? (4) If nature is culture, is culture nature? (5) How can consciousness, through conscience, help us heal our alienation from our home? and (6) Is the exploitation of our planet aligned with our exploitation of human minorities?

Questions 2 to 5 address the nature–technoculture binary that we have tackled thus far. While we will address the first question later in this chapter, the sixth and last question is particularly significant in understanding the relationship between the two gardens in *Mononoke Hime*. Gifford traces the idea that "challenging the exclusion of particular humans from equal standing in global society […] goes hand in hand with challenging a system that imperils all earthly beings" (Theriault 2015) back to ecofeminist authors like Carolyn Merchant (Gifford 2012, 26). The main actors in *Mononoke Hime* represent different types of struggles against oppressive military, patriarchal and exploitative forces: the women and the sick in Tatara, the endangered forest and its denizens, and Ashitaka, whose village is the last surviving pocket of the Emishi, an indigenous minority whose resistance against the expansionism of the Yamato Japanese challenges the myth of monoethnic Japan (Friday 1997, 1, 4). The fact that Eboshi's determination to defend the people of Tatara results in a war against the forest gods tells of the insidious ways in which oppression can turn vulnerable groups on each other. Nevertheless, the issue in *Mononoke Hime* runs deeper than the victimization of Eboshi and Tatara as justification for morally wrong deeds. Rather, such layering of conflicting interests from different parties whose actions cannot be dismissed as solely right or wrong, legitimate or illegitimate, is brought together by the movie's finale, pointing towards the need for a Safe Operating Space more than any "simplistic moral equation of industrial equals evil" (Napier 2005, 246).

Negotiating a Safe Operating Space

The concept of Safe Operating Space emerged in the Earth Systems (ES) sciences, namely from the work of Johan Rockström, Will Steffen and colleagues (Rockström *et al.* 2009; Steffen *et al.* 2015) on planetary boundaries. This framework defines a set of global biophysical indicators (e.g. climate change, land-system change, freshwater use, biosphere integrity) and attributes to each of them a boundary, a quantifiable limit. According to these scientists, overshot boundaries risk pushing the Earth System beyond "a relatively stable, 11,700-year-long Holocene epoch, the only state of the planet that we know for certain can support contemporary human societies" (Steffen *et al.* 2015). The fact that some of these boundaries have already been overshot due to human activities is one of the main arguments in favour of the adoption of the Anthropocene as our current geological epoch. The planetary boundaries framework combines the scientific understanding of how Earth Systems function with a precautionary approach[9] to "identify[ing] levels of anthropogenic perturbations below which the risk of destabilization of the ES is likely to remain low" (Steffen *et al.* 2015). In other words, it defines a Safe Operating Space in which human societies can continue to develop and thrive. This barometer, however, is based solely on biophysical boundaries, raising tensions and anxieties around who defines, safeguards and benefits from it. For instance, Sverker Sörlin, who contributed to the planetary boundaries articles, admits that the concept is useful, but does not shy away from criticizing it. For Sörlin, it is a "spectacular attempt" to keep the environment exclusively as an area of quantitative scientific expertise, "part of a larger pattern, which keeps at a distance value-based concerns and social and human aspects that otherwise seem crucially important" (Sörlin 2013). Like a techno-garden, the Safe Operating Space is limited by defined boundaries and quantifiable indicators.

The Safe Operating Space, then, is a useful if problematic concept. Christophe Bonneuil and Jean-Baptiste Fressoz (2016, 88) warn us about this "geopower that establishes the Earth as a system to know and govern as a totality, in all its components and all its functions". But if these planetary boundaries become more than biophysical indicators "held at a distance" (Barad 2007, 89), the Safe Operating Space can serve as an organizing principle for co-existing in an increasingly complex and plural world. In their appeal for the Humanities and the Social Sciences to enter the Anthropocene debate, Palsson and colleagues proposed something similar: "Planetary limits and boundaries' ideas need to incorporate human experience and must be sensitive to context and to the nature constructed by humans, embedded in a framework that includes issues of equity and environmental effects on humans" (Palsson *et al.* 2013, 11). One way to do this is to search for what a Safe Operating Space might look like in various places – both real and speculative. In *Mononoke Hime*'s case, the Safe Operating Space connects to what Saeki Junko argues is the main concern of the movie; that is, "to promote a willingness to accept difference as an essential part of life" (Napier 2005, 237).

Indeed, *Mononoke Hime* ends in an apocalypse caused by the breaking of environmental boundaries. As a result of the final confrontation between Eboshi and the forest's denizens, the *shishi-gami* becomes an amorphous death god, looking for its head after it was severed by Eboshi and stolen by one of the Emperor's lackeys. Everything it touches decays and dies, bringing about the destruction of both the forest and Tatara. Eventually, the *shishi-gami* perishes along with Moro and other gods, and even Eboshi is badly injured, her arm brutally bitten off by the wolf goddess. In extremis, Ashitaka and San manage to chase the Emperor's lackey and successfully return the severed head to the *shishi-gami*, preventing it from dying in agony. As the *shishi-gami* finally collapses and disappears, its last breath rewards them by covering the surrounding hills in fresh grass, lifting Ashitaka's curse and curing the lepers – an action, again, linking the *shishi-gami* to Eboshi, as figures capable of compassion and destruction alike. Despite the rebirth symbolized by the fresh grass, there is a sense of irreparable loss. This is somewhat compensated by both sides coming to a (perhaps temporary?) truce, represented by Ashitaka and San's agreement to live apart but to visit each other. "I like you, Ashitaka, but I can't forgive the humans",[10] says San; to which Ashitaka replies, "That's fine. You live in the forest, and I'll live in Tatara. Let's live, together. I'll come visit you, riding Yakul."[11] This unique resolution, evading the standard hero reward in which a male warrior gets the girl and a kingdom to rule, is reinforced by the final interventions of various characters. For instance, one villager of Tatara blurts out, "Eh, who would have thought the *shishi-gami* was the Hanasaka Jijii"[12] while gazing in wonder at the city's ruins covered in grass and flowers. Ashitaka tells San that the *shishi-gami* cannot die because it is life itself, and that by breathing life into the site of destruction it is telling humans and non-humans to survive. Eboshi herself admits that she "survived by riding on the back of a wolf",[13] acknowledging the wolves and San's vital role in saving her and the people of Tatara from the *shishi-gami*'s wrath. Yet, as she reassures the people of Tatara that they will start over and make "a good village" (*ii mura*), the exact meaning of her words is ambivalent. Is it a greener, more sustainable village or a larger, more intensive one? All in all, the give-and-take ending of Princess Mononoke disavows any simplistic polarization of the conflict between nature and technoculture, focusing instead on the intra-actions between opposing sides representing "real material differences but without absolute separation" (Barad 2007, 89).

That's all right. You live in the
forest, and I'll live at the ironworks.

Figure 12.5 San and Ashitaka's agreement at the end of *Mononoke Hime*. Video still from
Hayao Miyazaki's *Mononoke Hime* directed by Hayao Miyazaki and produced
at Studio Ghibli, Inc.
Source: © 1997 Nibariki TNDG.

The end of *Mononoke Hime* hints at a tense Safe Operating Space. The ambiguity of Eboshi's words, the fragile truce and the spectre of the military and power elites – who may have lost a battle but are set to return as soon as the ransacking or exploitation of Tatara becomes profitable again – point to the realization that instead of offering "solutions that will make everyone safe and happy [...] *Mononoke-hime* promises [...] that every day offers a new knife's-edge moral choice" (Abbey 2015, 118). In this way, Miyazaki destabilizes viewers' expectations while staging the post-pastoral question of "whether awe in attention to nature can lead to humility, reducing the hubris of human societies?" It seems that, to Miyazaki, the real measure of progress is a social emancipation without its modern tether of emancipation from nature. Instead of thinking about social emancipation solely in terms of an identity politics-based reform to a system that externalizes ecological phenomena, *Mononoke Hime* echoes the ecofeminist search for a cosmopolitics that overcomes, or at least mitigates, binary thinking, in particular, the power dynamics dictated by a "master subject" (Plumwood 1994, 42) over the subaltern Other. As authors such as Carolyn Merchant (1990), Val Plumwood (1994) and Kate Soper (1995) point out, these "others" are permutable among nature, non-humans, women and minorities. Any solution, then, must erode dualism via the affirmation of difference without superiority or mastery. *Mononoke Hime* reads as a cautionary warning to any viewpoints attempting to resolve our current predicaments by looking for manicheistic solutions, be they anthropocentric or biocentric, techno-optimist or techno-phobic.

Some discourses on the Anthropocene manifest this mindset of mastery; for instance, the Ecomodernist school of thought, comprising mainly natural scientists and engineers, caters to governments and business elites with its entrepreneurial discourse (Bonneuil and Fressoz 2016). For ecomodernists, the Anthropocene is "ripe with human-directed opportunity" (Ellis 2012), as humans can become gardeners of the planet, administering the – externalized – environment through geoengineering. History, however, shows us that the unintended perils of technological and ecological transformations grow in proportion to the extent and intensiveness of the efforts to impose order over the non-human world (Krishnan, Pastore and Temple 2015). The metaphor of the Garden Earth (i.e. that we should treat the planet as a garden) and the good Anthropocene are both co-optable discourses that can fuel our hubristic tendency to absorb nature into the sphere of human control and commodification, and therefore must be employed with caution, lest we end up with an unruly (Krishnan, Pastore and Temple 2015) planet.[14] The core "lesson" in *Mononoke Hime* is its appeal for humility in the face of alterity, both of social groups and nature. It asks for the recognition of the complexity and unpredictability of social-ecological phenomena that do not respond as linearly as neoliberal market ideologies would hope. Rather, the compromises, *rapprochements*, dissimilarities and ambivalences present in *Mononoke Hime* indicate a "narrative that relate[s] a dialogue with the Other" as an attempt at some kind of "global standard in a period of [...] internationalization, in which countries continue to maintain their identity while accepting the inevitable need for exchange with the Other" (Napier 2005, 246). The Safe Operating Space in *Mononoke Hime* is a constantly moving target, a perpetual negotiation between those who struggle to improve their condition and the agencies, both human and non-human, that consciously or inadvertently erode its boundaries.

Conclusion

To live in the Anthropocene is to know that we live precariously. It has always been so, since Earth dwellers have never been safe on a 30 km-thick crust floating over 2,900 km of magma. The question, then, is how to inhabit the Anthropocene – how to live (and die) well (Haraway 2015, 2016) in an era of planetary shifts and cultural clashes brought about by global capitalism and its increased mobilization of the Earth's inhabitants and resources. The challenge, for human societies in the Anthropocene, is how to safely inhabit spaces and times that are culturally and biologically diverse. Recently, this challenge has become steeper. The euphemisms post-truth and post-facts illustrate a trend by government officials, right-wing conservatism and corporations to dismiss the warning of environmental sciences. The rising mass discontent with globalization feeds a resurgence of nationalist agendas around the world, making the Other (the immigrant, the refugee, women, sexual minorities, non-humans, etc.) an increasingly marginalized subject that needs to be contained. At the same time, corporations and the free market are granted more and more liberties to extract and exploit resources and peoples.

Faced with this conundrum, Donna Haraway suggests that we need "stories [...] just big enough to gather up the complexities and keep the edges open and greedy for surprising new and old connections" (Haraway 2015, 160). Big enough to tackle our new planetary condition. Several years before the "cenes" (Anthropocene, Capitalocene, Chthulucene, etc.) were first mentioned, and many more before their rise in popularity, *Mononoke Hime*, with its challenging plot and spectacular visuals, made a very nuanced and idiosyncratic contribution to these debates. This explains its longevity as one of Hayao Miyazaki's most accomplished and popular films, both domestically and internationally, and why it remains, through the years, the subject of many articles and analyses. This chapter has looked at two locations in *Mononoke Hime* where the nature–culture dichotomy is strongly problematized: the film's two "gardens" of Tatara and the *shishi-gami's* pool. As the heart of the conflicting sides in *Mononoke Hime*, these gardens point to Miyazaki's troubling but not erasure of the nature–technoculture binary. Indeed, as Thomas Lamarre (2009, 306) points out, there is a lingering or mitigated Cartesianism in Miyazaki's use of depth that can be extrapolated to his particular brand of post-pastoralism. The gardens in *Mononoke Hime* are a "place apart" as much as within, in which viewers are made to experience the enduring, archetypal stickiness of distinctions whose boundaries are no longer fixed or absolute. Perhaps because the characters in *Mononoke Hime* are not free, and their world is not egalitarian, Miyazaki cannot envision a solution for human–non-human relations through the harmonious erasure of binaries. Instead, he focuses on *conflicts* arising from the demarcation of troubled gardens, including their dissolution.

Likewise, *Mononoke Hime* challenges the exploitation of the Other by hegemonic powers, but the Other is not abstracted as an uncomplicated, monolithic entity who can do no wrong. Women, indigenous peoples, the planet itself, have their inner power dynamics and agendas. Their empowerments are not interchangeable, and they do not always intersect or strive for the same emancipatory goals. What the gardens in *Mononoke Hime* and the many agencies that revolve around them do require is "a mode for integrating and questioning" such inquiries about our relation to environment and ourselves "in a holistic stretching of our notions of humanity" (Gifford 2012, 28). Miyazaki imagines a model that can redefine and guide the relations among humans and non-humans in the Anthropocene as a cluster of constantly negotiated, provisional Safe Operating Spaces. As argued in this chapter, the concept of Safe Operating Space is useful not despite but because of its troublesomeness. Searching for (and maintaining) a social-ecological "safe space" is, necessarily, a task filled with anxiety and difficulty, a permanent operation of balancing agential conflicts, flows and boundaries. But this troublesomeness is what nudges us, as Terry Gifford puts it, "into some ways of answering the most crucial question of our time: what is the right relationship by which people and planet can live together?" (Gifford 2012, 28). That the answer is as troubled and precarious as the gardens in the film is *Mononoke Hime*'s contribution to a growing body of fictional imaginings of liveable futures for everyone on Earth.

Acknowledgements

FCT MCTES – Project PTDC/IVC-HFC/6789/2014 – ANTHROPOLANDS – *Engineering the Anthropocene: The Role of Colonial Science, Technology and Medicine on Changing of the African Landscape.*
FCT MCTES – Project PEst-OE/HIS/UI0286/2014.

Notes

1 According to the Millennium Ecosystem Assessment (2005, 71), "This scenario depicts a globally connected world relying strongly on environmentally sound technology, using highly managed, often engineered, ecosystems to deliver ecosystem services, and taking a proactive approach to the management of ecosystems in an effort to avoid problems."

2 曇り無き眼で見定め，決める. *Kumorinaki me de misadame, kimeru* (Ghibli Script 1997, 22). All the film's dialogue in this chapter is translated by us from the official Japanese transcript of *Mononoke Hime.*

3 ここは，みな恐れて近寄らぬ私の庭だ. 秘密を知りたければ来なさい. *Koko wa, mina osorete chikayoranu watashi no niwa da. Himitsu o shiritakereba kinasai* (Ghibli Script 1997, 23).

4 Eboshi calls this riffle an *ishibiya* (literally, "stone fire arrow"), which situates the film's action during the early modern period, in the mid-sixteenth century, when firearms were first introduced into Japan from abroad (Udagawa 1996, 103).

5 お若い方，私も呪われた身ゆえ，あなたの怒りや悲しみはよくわかる. わかるが，どうかその人を殺さないでおくれ. その人はわしらを人として扱って下さった，たった一人のひとだ. わしらの病を恐れず，わしの腐った肉を洗い，布を巻いてくれた! *O-wakai kata, watashi mo norowa reta mi-yue, anata no ikari ya kanashimi wa yoku wakaru. Wakaru ga, dō ka sono hito o korosanaide okure. Sono hito wa washira o hitotoshite atsukatte kudasatta, tatta hitori no hito da. Washi-ra no yamai o osorezu, washi no kusatta niku o arai, nuno o maite kureta!* (Ghibli Script 1997, 24–25).

6 生きることは，まことに苦しく辛い！世を呪い，人を呪い，それでも生きたい! *Ikirukoto wa, makotoni kurushiku tsurai! Yo o noroi, hito o noroi, soredemo ikitai!* (Ghibli Script 1997, 25).

7 In an interview, Miyazaki talks about the importance of these complex events in moulding the morality of *Mononoke Hime.* For Miyazaki, the Gulf and the Yugoslav wars made it impossible to align oneself fully with either faction, a stance reflected in Ashitaka's attempt to save those he loves without taking sides. Miyazaki predicts that taking sides as an external observer will be increasingly difficult in the future of world conflicts (Miyazaki 2014a, 35).

8 The original Chinese word is *jiejing.* It was defined in the late Ming period gardening manual *Yuanye* ("The Garden Treatise" or "The Craft of Gardens", 1631) by Ji Cheng.

9 The precautionary approach, or precautionary principle, is a "guideline in environmental decision making" based on "taking preventive action in the face of uncertainty; shifting the burden of proof to the proponents of an activity; exploring a wide range of alternatives to possibly harmful actions; and increasing public participation in decision making" (Kriebel *et al.* 2001).

10 アシタカは好きだ，だが人間げんを許すことはできない. *Ashitaka ha suki da, daga ningen wo yurusu koto ha dekinai* (Ghibli Script 1997, 75).

11 それでもいい.サンは森で，私はタタラ場で暮らそう. 共に生きよう.会いに行くよ，ヤックルに乗って. *Soredemo ii. San ha mori de, watashi ha Tatara-ba de kurasou. Tomo ni ikiyou. Ai ni ikuyo, Yakkuru ni notte* (Ghibli Script 1997, 75).

12 へえ，シシ神は花咲はなさかじじいだったんだ […] *Ee, shishi-gami ha Hanasaka Jijii dattanda* […] (Ghibli Script 1997, 74). The Hanasaka Jijii is a character from a

Japanese folk-tale. It means, literally, "The Old Man Who Makes Flowers Bloom". The tale is about the spirit of a dog, murdered by an envious neighbour, who visits his master in his dreams. To thank the master for giving it a proper burial, the dog performs several miracles, including making flowers bloom from a dead cherry tree. The master performs the miracle in front of the feudal Lord, who decides to reward the old man as a token of appreciation (Freeman-Mitford 2005).

13 私が山犬の背中に乗って生き残ってしまった [...] *Watashi ga yamainu no senaka no note ikinokotte shimashita* [...] (Ghibli Script 1997, 75).

14 Unruly environments are "places difficult to control and categorize, whether choked with vegetation, submerged underwater, or encased in concrete". These are places of extreme environmental qualities (salinity, aridity, altitude, prone to intense and regular wildfires, etc.) that are often "a product of our own making" and thwart stability, public health, progress and financial gain. Ultimately they remind us of the "limits of environmental control in an era of technological and institutional hubris" and question the relegation of agency solely to human intentionality (Krishnan, Pastore and Temple 2015).

Bibliography

Abbey, Kristen L. 2015. "'See with Eyes Unclouded': Mononoke-Hime as the Tragedy of Modernity". *Resilience: A Journal of the Environmental Humanities* 2(3): 113–119. https://doi.org/10.5250/resilience.2.3.0113.

Barad, Karen. 2007. *Meeting the Universe Halfway: Quantum Physics and the Entanglement of Matter and Meaning*. Durham, NC: Duke University Press.

Bonneuil, Christophe and Jean-Baptiste Fressoz. 2016. *The Shock of the Anthropocene: The Earth, History and Us*. London; Brooklyn, NY: Verso.

Ellis, Erle. 2012. "The Planet of No Return – Human Resilience on an Artificial Earth". *The Breakthrough* (blog).https://thebreakthrough.org/index.php/journal/past-issues/issue-2/the-planet-of-no-return.

Freeman-Mitford, Algernon Bertram. 2005. *Tales of Old Japan: Folklore, Fairy Tales, Ghost Stories and Legends of the Samurai*. Mineola, NY: Dover Publications.

Friday, Karl F. 1997. "Pushing beyond the Pale: The Yamato Conquest of the Emishi and Northern Japan". *Journal of Japanese Studies* 23(1): 1–24. https://doi.org/10.2307/133122.

Ghibli Script. 1997. "もののけ姫/Princess Mononoke (1997)".

Gifford, Terry. 1999. *Pastoral*. London; New York: Routledge.

Gifford, Terry. 2012. "Pastoral, Antipastoral, and Postpastoral as Reading Strategies". www.terrygifford.co.uk/Pastoral%20reading.pdf.

Haraway, Donna. 2015. "Anthropocene, Capitalocene, Plantationocene, Chthulucene: Making Kin". *Environmental Humanities* 6: 159–165.

Haraway, Donna. 2016. *Staying with the Trouble: Making Kin in the Chthulucene*. Durham, NC: Duke University Press.

Helphand, Kenneth. 1997. "Defiant Gardens". *The Journal of Garden History* 17(2): 101–121. https://doi.org/10.1080/01445170.1997.10412542.

Jensen, Casper Bruun and Anders Blok. 2013. "Techno-animism in Japan: Shinto Cosmograms, Actor-network Theory, and the Enabling Powers of Non-human Agencies". *Theory, Culture and Society* 30(2): 84–115. https://doi.org/10.1177/0263276412456564.

Kriebel, D, J. Tickner, P. Epstein, J. Lemons, R. Levins, E.L. Loechler, M. Quinn, R. Rudel, T. Schettler, and M. Stoto. 2001. "The Precautionary Principle in Environmental Science". *Environmental Health Perspectives* 109(9): 871–876.

Krishnan, Siddhartha, Christopher Pastore and Samuel Temple. 2015. "Unruly Environments". *RCC Perspectives*, no. 3.

Kuitert, Wybe. 2015. "Borrowing Scenery and the Landscape That Lends – The Final Chapter of Yuanye". *Journal of Landscape Architecture* 10(2): 32–43. https://doi.org/10. 1080/18626033.2015.1058570.

Lamarre, Thomas. 2009. *The Anime Machine: A Media Theory of Animation*. Minneapolis: University Of Minnesota Press.

Marris, Emma. 2013. *The Rambunctious Garden: Saving Nature in a Post-wild World*. New York: Bloomsbury USA.

Mazur, Eric Michael. 2011. *Encyclopedia of Religion and Film*. Santa Barbara, CA; Oxford: ABC-CLIO.

Merchant, Carolyn. 1990. *The Death of Nature: Women, Ecology, and the Scientific Revolution*. New York: HarperOne.

Millennium Ecosystem Assessment. 2005. *Ecosystems and Human Well-being: Synthesis*. Washington, DC: World Resources Institute.

Miyazaki, Hayao. 2014a. *Turning Point, 1997–2008*. Translated by Beth Cary and Frederik L. Schodt. San Francisco, CA: VIZ Media LLC.

Miyazaki, Hayao. 2014b. *The Art of Princess Mononoke*. San Francisco, CA: VIZ Media LLC.

Napier, Susan J. 2005. *Anime from Akira to Howl's Moving Castle: Experiencing Contemporary Japanese Animation*. New York: St Martin's Griffin.

Palsson, Gisli, Bronislaw Szerszynski, Sverker Sörlin, John Marks, Bernard Avril, Carole Crumley, Heide Hackmann, *et al.* 2013. "Reconceptualizing the 'Anthropos' in the Anthropocene: Integrating the Social Sciences and Humanities in Global Environmental Change Research". *Environmental Science and Policy*, Special Issue: Responding to the Challenges of our Unstable Earth (RESCUE), 28 (April): 3–13. https://doi. org/10.1016/j.envsci.2012.11.004.

Plumwood, Val. 1994. *Feminism and the Mastery of Nature*. London: Routledge.

Rockström, Johan, Will Steffen, Kevin Noone, Åsa Persson, F. Stuart III Chapin, Eric Lambin, Timothy Lenton, *et al.* 2009. "Planetary Boundaries: Exploring the Safe Operating Space for Humanity". *Ecology and Society* 14(2). https://doi.org/10.5751/ES-03180-140232.

Soper, Kate. 1995. *What Is Nature? Culture, Politics and the Non-human* (1st edition). Oxford; Cambridge, MA: Wiley-Blackwell.

Sörlin, Sverker. 2013. "Reconfiguring Environmental Expertise". *Environmental Science and Policy* 28 (April): 14–24. https://doi.org/10.1016/j.envsci.2012.11.006.

Steffen, Will, Katherine Richardson, Johan Rockström, Sarah E. Cornell, Ingo Fetzer, Elena M. Bennett, Reinette Biggs, *et al.* 2015. "Planetary Boundaries: Guiding Human Development on a Changing Planet". *Science* 347(6223): 1259855. https://doi. org/10.1126/science.1259855.

Theriault, Noah. 2015. "Gendering the Anthropocene". *Inhabiting the Anthropocene* (blog), 20 May. https://inhabitingtheanthropocene.com/2015/05/20/gendering-the-anthropocene/.

Treib, Marc. 1989. "Reduction, Elaboration and Yūgen: The Garden of Saihō-Ji". *The Journal of Garden History* 9(2): 95–101. https://doi.org/10.1080/01445170.1989.104082 71.

Treib, Marc. 1998. "Formal Problems". *Studies in the History of Gardens and Designed Landscapes* 18(2): 71–92. https://doi.org/10.1080/14601176.1998.10435535.

Udagawa, Takehisa. 1996. "The Introduction and Spread of the Ishibiya Firearm at the Beginning of the Early Modern Period". *Bulletin of the National Museum of Japanese History*, 66: 103–123.

Index